Monitoring Rocky Shores

Monitoring Rocky Shores

Steven N. Murray
Richard F. Ambrose
Megan N. Dethier

UNIVERSITY OF CALIFORNIA PRESS
Berkeley Los Angeles London

University of California Press
Berkeley and Los Angeles, California

University of California Press, Ltd.
London, England

Library of Congress Cataloging-in-Publication Data

Murray, Steven N. (Steven Nelson), 1944–
Monitoring rocky shores / Steven N. Murray, Richard F. Ambrose, Megan N. Dethier.
p. cm.
Includes bibliographical references.
ISBN 0-520-24728-0 (cloth : alk. paper)
1. Intertidal ecology—Research—Methodology.
2. Environmental monitoring—Methodology. I. Ambrose, Richard F. (Richard Francis) II. Dethier, Megan Nichols.
III. Title.
QH541.5.S35M87 2006
577.69'9–dc22

2005011949

Manufactured in Canada

15 14 13 12 11 10 09 08 07 06
10 9 8 7 6 5 4 3 2 1

The paper used in this publication meets the minimum requirements of ANSI/NISO Z 39.48-1992 (R 1997) (*Permanence of Paper*).

Cover photo: Ochre sea stars feeding on mussels near the Bamfield Marine Station in Barkley Sound, British Columbia.

CONTENTS

LIST OF FIGURES

LIST OF TABLES

PREFACE

The purpose of this book is to provide interested investigators with the information needed to develop methods and procedures for carrying out key elements of a rocky intertidal field-sampling program. Critical discussions and evaluations of the various elements of an effective rocky intertidal field-sampling program are provided in the ensuing chapters. The book was written for research and agency scientists, agency managers, and advanced university students who might benefit from consolidated discussions and reviews of important sampling issues and field procedures for designing and evaluating field monitoring and impact studies performed on rocky intertidal macroinvertebrates, seaweeds, and seagrasses. Emphasis is placed on describing and discussing options for field methods and procedures, with a focus on their use in monitoring programs and impact studies. Users are required to formulate their own study goals and study designs. Clearly, any effective study program must have clear goals and objectives and invoke robust study designs.

In this book, we have attempted to break down the decision-making process into its various elements so investigators can become aware of the advantages and disadvantages of choosing a particular method or approach and can consider trade-offs between the time and cost and the value of the collected data. No book can serve as a menu of fixed procedures for sampling complex rocky intertidal environments. Spatial variability and different site histories mean that the methods and procedures that best match study goals at one site may not be optimal at another site where the habitat topography and the distributions and abundances of targeted species vary. For example, the size and number of sampling units required to effectively determine the abundances of a dense barnacle population at one site might differ from those selected to optimally sample the same species at another site where densities are much lower or where the habitat is more topographically heterogeneous. Hence, it is impossible to prescribe standard protocols or procedures that can be used effectively to sample rocky intertidal organisms under all encountered conditions. Instead, sound scientific studies can only be designed following careful consideration of each element of the field-sampling program. This requires that investigators have sufficient information to select procedures that best address study goals and questions given consideration of site-specific details, tidal schedules, and available resources. In the following chapters, we sequentially present a brief overview of monitoring and impact study designs (chapter 1) and discussions of factors involved in site selection (chapter 2), biological units to be sampled (chapter 3), field sampling layouts and designs (chapter 4), selection of sampling units (chapter 5), nondestructive (chapter 6) and destructive (chapter 7) methods of quantifying abundance, and methods for measuring age, growth rates, size structure, and reproductive condition of intertidal organisms (chapter 8).

ACKNOWLEDGMENTS

Putting together this book proved to be a much longer and more difficult task than originally envisioned, and we would not have been able to complete this project without the contributions, assistance, and patience of many individuals.

First, our thanks go to Fred Piltz, Mary Elaine Dunaway, and Maurice Hill of the Minerals Management Service (MMS; Pacific Region) of the U.S. Department of the Interior for their assistance and support throughout this project. It is clearly to their credit that Fred, Mary Elaine, Maurice, and others at the MMS recognize that strong and well-thought-out study and sampling programs cannot be taken for granted, particularly in variable rocky intertidal habitats. We thank Mary Elaine for her support, encouragement, and patience throughout this project.

Funds to organize and develop this handbook were provided by the MMS through the Southern California Educational Initiative (SCEI) program (MMS Agreement 14-35-0001-30761) administered by the Marine Science Institute (MSI) of the University of California, Santa Barbara. We are grateful for the patience and interest in this project displayed by Russ Schmitt, Director of the SCEI program, and very much appreciate the excellent administrative support provided by Bonnie Williamson of the MSI, who served as SCEI Program Manager.

Mary Elaine Dunaway and Maurice Hill of the MMS, John Tarpley and Rob Ricker of the Office of Spill Prevention and Response (OSPR) unit of the California Department of Fish and Game, and John Cubit of NOAA's Natural Resources Damage Assessment (NRDA) Division provided guidance and direction in the initial stages of developing this handbook, with the hope that multiple needs could be met by a single product. We are grateful for the guidance, criticism, and insight provided by Mary Elaine, Maurice, John, Rob, and John. Their reviews of drafts of this handbook are much appreciated.

Peer reviews of the chapters in this handbook were carried out by several individuals besides our MMS, OSPR, and NOAA colleagues. Specifically, we thank Gray Williams, Susan Brawley, Roger Seapy, the late Mia Tegner, Pete Raimondi, Jack Engle, Helen Berry, and Lucinda Tear for reviewing whole or sections of chapters. Numerous other individuals were also contacted for advice and discussion during development of this work. We are grateful for their comments and insight and the many constructive suggestions made by our reviewers. The assistance of our reviewers and colleagues clearly made the final draft of this book a better product. We take responsibility for any remaining mistakes, errors, or inaccuracies, and appreciate the efforts of our reviewers to keep us on the right path.

Finally, we thank Esther Seale for clerical support, Kelly Donovan for assistance with most of the line art, Janine Kido for editorial assistance, and Jayson Smith for support during various stages of this project. S.N.M. would like to acknowledge Phyllis Grifman of the University of Southern California (USC) Sea Grant Program for encouraging work on this handbook and the National Sea Grant Program and California State Resources Agency (NOAA Grant NA46RG0472 to the USC Sea Grant Program) for funding studies on rocky intertidal systems that contributed greatly to the development of the discussions in chapters 1, 6, 7, and 8 of this handbook. S.N.M. would like to thank C. Eugene Jones (Chairperson) and the Department of Biological Science at California State University, Fullerton, for support and assistance during the development and completion of this project. Thanks also are due to Shani Bergen Murray for providing suggestions that improved the consistency of our writing and to Cheryl Aranda who helped with indexing. R.F.A. acknowledges the California Coastal Commission for funding studies that contributed to the discussions in chapters 4 and 5. M.N.D. gratefully acknowledges helpful discussions with and input from Carl Schoch and Lucinda Tear, and support from the Friday Harbor Laboratories during the writing of chapters 2 and 3.

The views, opinions, and conclusions in this book are solely those of the authors and do not necessarily reflect the views or official policies, either expressed or implied, of the MMS, OSPR, or NOAA or any other federal, state, or local government or governmental agency of the United States. Mention of trade names or commercial products does not constitute an endorsement or recommendation for use. This work was reviewed and approved for publication by the Pacific Outer Continental Shelf Region, MMS, and U.S. Department of the Interior. This handbook has not been edited for conformity with MMS editorial standards.

Rocky intertidal bench at Victoria Street Beach, Laguna Beach, California.

CHAPTER 1

Designing Rocky Intertidal Monitoring and Impact Field Studies

A Brief Overview

INTRODUCTION

Field sampling programs provide the information needed to determine the status and dynamics of populations and communities and thus are the foundation for many kinds of research, including impact and monitoring studies. Certain habitat types present more formidable challenges compared with others when it comes to designing and performing field studies. The rocky intertidal zone is one of these habitat types. Basic descriptions of the natural history and abiotic and biotic patterns and processes that characterize rocky intertidal environments are not addressed here. The interested reader can find such discussions in many sources, including, for example, Lewis (1964), Stephenson and Stephenson (1972), Carefoot (1977), Kozloff (1983), Ricketts et al. (1985), Raffaelli and Hawkins (1996), Knox (2000), and Menge and Branch (2000). The unique features of rocky intertidal environments need to be understood in order to select the most appropriate field procedures for sampling rocky intertidal organisms.

The physical and biological complexity of most rocky shores results in high variability in almost any measured parameter, even over very short vertical and horizontal scales. The sources of this variability must be taken into account when designing and performing field sampling programs in rocky intertidal habitats.

Unlike terrestrial habitats, the intertidal zone is accessible for most studies only during limited periods when the tide is low. These periods will vary daily in both tidal magnitude and the time of day when they occur. Hence, intertidal investigators must develop work calendars and plans dictated by the tides and be prepared to perform fieldwork during almost any hour of the day depending on tidal schedules. In addition, daily fieldwork must be performed efficiently in order to complete sampling programs during the few hours when the tide is out and the shore is accessible.

Although environmental conditions are usually quite constant in most marine habitats, the intertidal zone experiences large changes over the daily tidal cycle, with higher shore elevations receiving longer periods of emersion and exposure to physically harsh conditions. This creates a strong vertical gradient in stressful physical conditions running up and down the shore, which contributes to the well-known zonation patterns displayed by rocky intertidal organisms. Rocky shores also show horizontal gradients of exposure to wave action and, if near freshwater inputs, salinity. In addition, rocky shores vary over horizontal scales in substratum type and topography. Some shores are dominated by expansive flattened benches, others by steeply sloped vertical stacks, while others consist mostly of boulders and rocks that can vary in size from centimeters (shingles, cobbles, and small boulders) to meters (large boulders). Crevices, channels, and pools typically break up even the most flattened rocky intertidal benches. Thus, almost all rocky intertidal habitats appear as a heterogeneous mosaic of microhabitat types resulting in a landscape that varies greatly even over very short vertical and horizontal distances.

Rocky intertidal habitats are perhaps best known for the rich variety of organisms that they support. Although fish and other mobile organisms play a role during periods of tidal immersion, slow-moving and sessile invertebrates, seaweeds, and seagrasses dominate the intertidal landscape during periods of low tide, when rocky intertidal habitats can most easily be accessed. Historically, it is this rich and diverse assemblage of benthic invertebrates and marine plants, distributed over short expanses of variable shoreline, that has attracted the public, educators, and members of the scientific community to engage in exploration and study of the rocky intertidal zone. The strong vertical and horizontal gradients in environmental conditions affect the distributions and abundances of these organisms and contribute to community structures that vary with vertical shore position, from sheltered to exposed conditions, across salinity gradients, and with substratum type and configuration. Because of the high phyletic diversity encountered on most rocky shores, investigators performing community-level studies must have extensive and broad taxonomic expertise.

Species distributions and abundances not only vary over spatial scales but also can show considerable temporal variation over time scales ranging from years to decades, even in the absence of major disturbances. Knowledge of the historical patterns of physical and biological disturbance, recruitment, and other processes that might drive temporal variation in rocky intertidal populations and communities is almost never available. This means that habitat histories can vary considerably over spatial scales ranging from kilometers to meters, and significantly influence species distributions, abundances, and population structures.

Most methods and procedures for designing and performing field sampling programs can be readily transferred among habitat types. For example, there is a long history of testing and evaluating field procedures for sampling vegetation and animals in terrestrial environments, and many of these procedures have been adopted to sample rocky intertidal organisms. However, the high spatial variability in both abiotic and biotic features typically encountered in rocky intertidal sampling programs usually exceeds variability at the physical and biological scales addressed by most terrestrial sampling programs. Established methods and procedures need to be carefully analyzed and selected so as take into account the high "noise-to-signal" ratios almost surely to be encountered in rocky intertidal habitats when performing quantitative sampling programs. For these reasons, much discussion of rocky intertidal sampling approaches and methodologies continues to take place. The purpose of this handbook is to provide interested investigators with the information needed to determine the most appropriate methods and procedures for carrying out key elements of a rocky intertidal field-sampling program.

DESIGN ELEMENTS FOR ROCKY INTERTIDAL FIELD STUDIES

The steps in designing a field sampling program include (1) identifying the study goals including the questions to be answered by the study or the hypotheses to be tested and (2) developing an effective and statistically powerful study design. These steps are followed by determining study sites, the biological units to be sampled, the sampling design, layout, and units to be employed, and the type of data to be obtained (fig. 1.1). Identifying study goals and developing effective and appropriate study designs are discussed only briefly here. Emphasis in this handbook is placed on the methods and procedures for carrying out field monitoring, impact, or other ecological studies once the goals and overall design have been determined.

Study Types and Goals

The specific goals and objectives of field studies must be identified and will dictate the overall study design and elements of the field-sampling

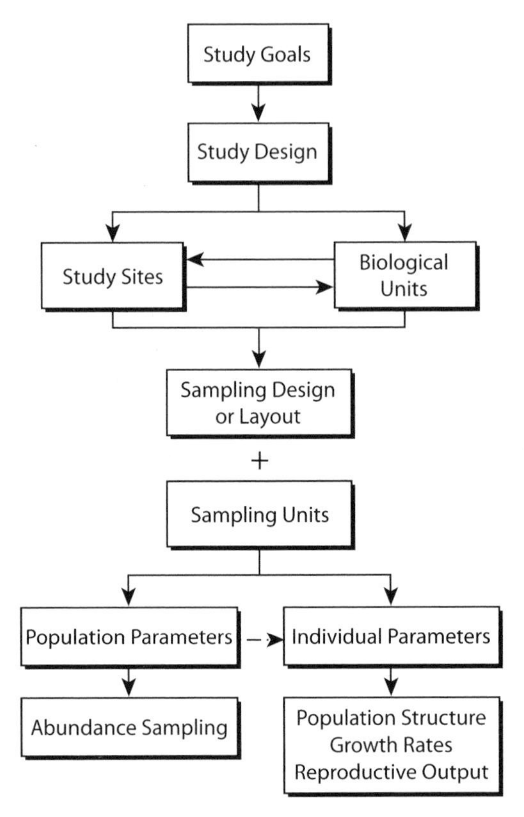

Figure 1.1. Decision tree for designing rocky intertidal sampling programs. The goals of most monitoring or impact sampling programs will focus on either specific sites and habitat types or biological units. Decision flow in these studies should be dictated by program goals. Selection of sampling units and sampling design or layout should be selected in coordination to match biological units and the types of data (population-based or individual-based parameters) to be collected.

program. Maher et al. (1994) stress the need for clearly stating the problem of concern or reason for sampling and identifying the study objectives, which can then be stated as testable hypotheses. Most ecology texts outline the basic elements of developing study questions or hypotheses; Kingsford and Battershill (1998) provides a particularly good discussion of procedures for establishing field studies. Field studies can generally be classified into four general categories (Kingsford et al. 1998). These are (1) baseline studies, (2) impact studies, (3) monitoring studies, and (4) patterns and processes (ecological) studies. Using their definitions as a foundation, these study types are defined herein as follows. Baseline studies are considered to have the goal of determining the present status of biological conditions, for example, species abundances or community composition. These often are referred to as one-time or "one-off" studies because they are planned to be executed at only a single point in time. The goals of impact studies include determining and relating biological changes to a particular known perturbation, such as a sewage outfall or an oil spill, and also measuring the spatial and temporal scales of detected changes. Monitoring studies focus on the following biological parameters through time, usually with the objective of detecting changes in measured parameters due to existing or future threats from human activities. Studies of patterns and processes are types of ecological field assessments that describe species distributions and abundances (patterns) with the intent of determining the factors that cause them (processes) (Kingsford et al. 1998). Process studies generally involve both descriptive and experimental approaches. Although many of the same considerations apply to the actual execution of patterns and processes studies, emphasis in this handbook is placed on rocky intertidal baseline, impact, and monitoring field studies. Discussions of patterns and processes studies can be found in works by Underwood (1993a, 1997) and many other sources and are not covered here.

Study Designs

Once study goals are clearly identified, the next step in developing a field-sampling program is to determine an effective study design. The specifics of this design will vary with the study type and goals, but also will be dictated by circumstances such as the availability of suitable study areas and available resources. Regardless of the goals of the field study, Kingsford and Battershill (1998) describes considerations that should be incorporated into the study design. These include (1) the need for controls in both space and time, (2) predetermination of the methods of data analysis required to answer study questions, (3) replication of every level of sampling, (4) the use of multiple locations for sampling, (5) potential effects of existing variation over short temporal and small spatial scales, (6) insurance that independent and replicate samples are to be taken, (7) depending on the design, that sampling is done randomly (preferably) or haphazardly, and (8) that quantitative results can be expressed with a measure of statistical variability. Most rocky intertidal field sampling programs will address questions that focus either on study sites or on biological units (e.g., targeted populations or communities) depending on the study goals.

One-Time or "One-Off" Baseline Studies. These studies are generally performed to obtain information about the present status of populations or communities. Because, by definition, they are not replicated in time, they will be of only limited value. Results can be used to evaluate the "health" of an intertidal habitat in the absence of historical data by comparing measured parameters with information obtained in the literature or from other study areas. Data also can be used to produce a record of baseline conditions prior to proceeding with, for example, a coastal development or establishment of a marine protected area and to predict the impact of such activities (e.g., risk assessment *sensu* Suter 1993). However, one-time baseline studies in rocky intertidal habitats will only be useful for detecting the grossest changes in populations and communities (Hawkins and Hartnoll 1983) or where the biological outcomes of identified stressors or events are well understood. Another implicit goal of one-time studies is often to establish a baseline for the biological populations or communities against which change can be measured at some unknown future time. The lack of replication through time in one-off studies means that there is no chance to quantify natural variability in biological parameters and, thus, to distinguish future changes resulting from a particular event or stressor from those reflecting natural temporal variation. Baseline data, however, might allow retrospective risk assessment. If the potential for future resampling is envisioned, specific site locations (e.g., transect heads or reference points) should be thoroughly documented to allow relocation, and robust and repeatable sampling procedures should be employed when performing a one-time study. Kingsford and Battershill (1998) also emphasize the advantages of sampling multiple locations within the defined study area when performing one-time baseline studies.

Impact Studies. Impact studies are designed to determine the changes brought about by a perturbation or stressor by comparing the status of natural or unimpacted biological or other measured parameters with their status under impacted conditions. Impact study designs have received much attention since Green's (1979) pivotal treatment of impact sampling designs and statistical methods. Excellent discussions of the advantages and disadvantages of various impact study designs are given by Stewart-Oaten et al. (1986), Underwood (1991, 1992, 1993b, 1994), Osenberg and Schmitt (1996), Osenberg et al. (1996), Stewart-Oaten (1996), Ellis and Schneider (1997), Kingsford and Battershill (1998) and Kingsford (1998). Researchers interested in determining impacts of human activities or natural events on rocky intertidal populations and communities should become well informed about the advantages and disadvantages of the various impact study models before designing a field sampling program.

Several models generally are used in performing field studies of impacts (fig. 1.2). Unfortunately, choice of models is often dictated by circumstances such as the availability of sites, funding levels, and timing, instead of the most ecologically robust study design. The most robust models, particularly for sampling rocky intertidal environments that show high biological variation in both space and time, require the collection of data prior to the onset of an impact. This means that a commitment of funds and implementation of a scientifically sound study design must proceed well in advance of an impact, events that can only occur in the case of planned coastal developments or activities. For unplanned impacts, such as a major oil spill, robust before-impact data sets will almost always be absent unless a commitment has been made to carry out a long-term and scientifically sound, regional monitoring program. In southeastern Alaska, the lack of robust data sets prior to the *Exxon Valdez* oil spill was identified as a major problem in designing effective postspill impact studies (Paine et al. 1996).

Gradient models (fig. 1.2A), where sampling is done at intervals distributed at variable distances from the source of impact, have long been used to detect changes in biological parameters. ***Gradient impact designs*** are particularly useful where a stressor or disturbance attenuates with distance from a point source, whereas other designs are more appropriate where impacts have distinct boundaries and control and impact sites can be identified (Ellis and Schneider 1997). Examples where gradient designs are appropriate include cases where an impact is due to discharge of a sewage effluent or an oil drilling operation. In these cases, field study sites can be set up in a grid both within and at various distances from the point source of the impact. Analyses of quantitative data collected in such a gradient design can be carried out by regression or analysis of covariance (ANCOVA) or by using multivariate techniques. A gradient sampling design was found to be more powerful than other design alternatives in detecting certain effects of oil drilling operations in the Ekofisk oil field in the North Sea (Ellis and Schneider 1997).

Several impact study designs are appropriate where distinct control and impact conditions can be determined in time or space. One such model (fig. 1.2B) involves collecting data at a single site both before and after an impact. Impact determination is based on finding significant statistical differences between parameters measured before and after the impact. This ***before–after design*** avoids problems of spatial variation, an important consideration in rocky intertidal sampling programs, by restricting sampling to the same site, with samples being taken at one or

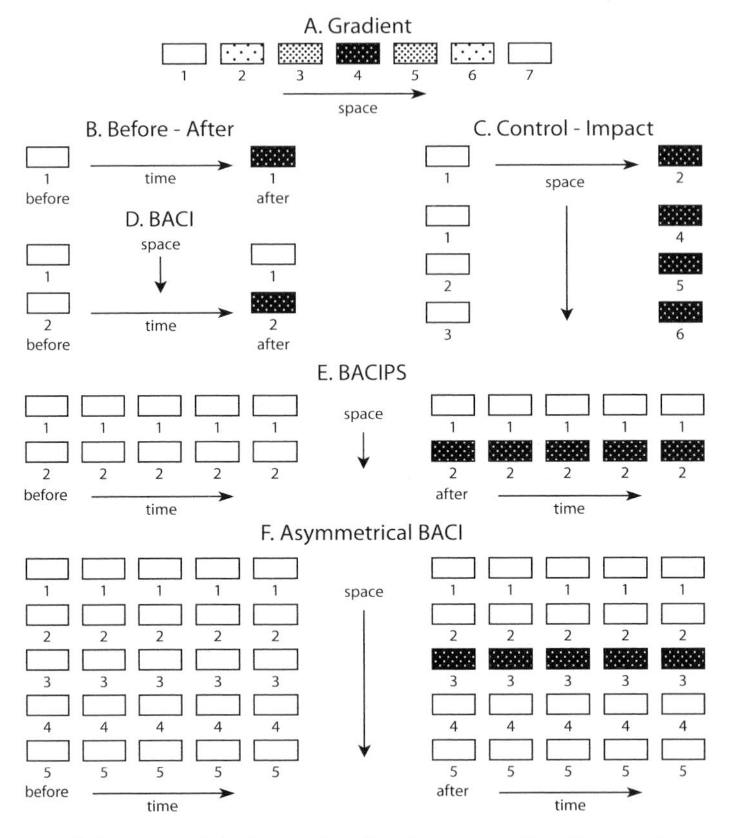

Figure 1.2. Selected models for performing impact studies. Rectangles represent sites and individual sites are numbered; degree of shading represents magnitude of impact. (A). **Gradient design** where sites are distributed at various distances from a central impact site, creating a gradient of control and impact conditions. (B) **Before–after design,** where the same site is studied both before and after an impact. (C) **Control–impact design,** where data are collected at a single point in time from impact site or sites and compared with data collected from control site or sites. (D) **BACI design,** where data are collected from both control and impact sites both before and after an impact. (E) **BACIPS design,** where data are collected on multiple occasions (here, five times) from control and impact sites both before and after an impact. (F) **Asymmetrical BACI design,** where data are collected on multiple occasions (here, five times) both before and after an impact, usually from one impact and multiple (here, four) control sites. These different study designs have advantages and disadvantages and require different statistical methods for analyzing the collected data. However, all designs require decisions on site selection, biological units, sampling units to be employed, the design for locating sampling units, and the type of data to be taken.

more times before and one or more times after an impact. Although eliminating the potentially confounding effects of spatial variation, this impact study design is seriously flawed by its failure to separate natural temporal variability from effects due to the impact. In other words, this design cannot eliminate the possibility that detected changes have resulted from other regional events (e.g., increased sea temperatures) instead of the impacting agent.

Another often-used model in impact studies is the ***control–impact design*** (fig. 1.2C), where studies are performed at one or more control and one or more impact sites. This is a commonly used study design, probably because it does not depend on before-impact data, which usually are unavailable. Although accounting for temporal variability by sampling both control and impact sites at essentially the same time, this design is unable to separate natural spatial variability from effects due to the activity (see chapter 2). The design is based on the assumption that control and impact sites are identical except for the impact, an occurrence that is highly unlikely particularly in complex and dynamic rocky intertidal habitats. Greater ability to statistically discriminate impacts from natural spatial variation among study sites can be obtained by replicating control and impact sites. For example, nested analysis of variance (ANOVA) can be used in selected cases to discriminate among impacted and control (unimpacted) sites (McKone and Lively 1993). However, in rocky intertidal and other spatially variable habitats, differences within replicated site parameters often swamp differences between control and impact site groups for all but the largest impact effects. In most cases, for example, point source discharges from a sewage or power plant outfall, impacted sites cannot be replicated, leaving a choice of comparing one or more control sites with a single impacted site.

A fourth design addresses the problems of spatial and temporal variation by combining the two designs into a single design—what Green (1979) referred to as the ***before–after–control–impact (BACI) design*** (fig. 1.2D). There are several ways to employ this type of design depending on the number of sampling periods and sites to be sampled. In the simplest case, one control and one impacted site are sampled one time before and one time after the impact. As described by Osenberg and Schmitt (1996), impact determination is then based on the "interaction between Time and Location effects, using variability among samples taken within a site (on a single date) as the error term." Unfortunately, as illustrated by Osenberg and Schmitt (1996), successful use of this design depends on the variables measured at the control and impact sites following the same trajectories through time; a significant interaction only shows that the measured variables did not track one another before and after the impact.

As pointed out by Hurlbert (1984), this circumstance may or may not be attributable to the impact perturbation.

To correct deficiencies in the BACI design, Stewart-Oaten et al. (1986) argued for a design based on comparisons of a time series of differences between control and impact sites taken before and after an impact. This design (fig. 1.2E) has been referred to as the ***before–after–control–impact paired series (BACIPS) design*** (*sensu* Osenberg and Schmitt 1996). Impact determination in this design is based on statistical comparison of the before differences with the after differences for the variables measured at the two sites. The assumption in this analysis is that each difference in the before period is an independent estimate of the natural spatial variation between the control and the to-be-impacted sites. As discussed by Osenberg and Schmitt (1996), this design also has limitations, including that it lacks spatial replication (Underwood 1994), but these are clearly understood and have been delineated by Stewart-Oaten et al. (1986) and Stewart-Oaten (1996). In the latter contribution, Stewart-Oaten discusses additional analysis options that strengthen the efficacy of the BACIPS Design.

Underwood (1991, 1992, 1994) discusses an impact study design that overcomes deficiencies in the previous BACI-based models, including the BACIPS model. This is the ***asymmetrical before–after–control–impact (BACI) design*** (fig. 1.2F). In this design, multiple controls and a single impacted site are studied both before and after an impact. In this design, impact determination is made by observing statistical differences between changes at multiple (and presumably variable) control sites and changes at the impacted site using analysis of variance (ANOVA). The use of multiple controls overcomes several problems with BACI designs, including the problem of spatial–temporal interactions in BACIPS designs.

Monitoring Programs. Monitoring programs by definition involve the repeated sampling of measured parameters over time. Monitoring programs in aquatic environments have focused on chemical, physical, and biological parameters, and have addressed goals as diverse as detecting microbial contamination of beaches, determining the concentrations of potentially harmful materials in fish, and learning whether the effects of discharged sewage are creating biological changes across the sea floor. Rocky intertidal habitats are important candidates for establishing effective monitoring programs because of their high public visibility and value and their potential to undergo degradation from human activities.

Karr and Chu (1997) recently reviewed the history of monitoring programs in aquatic systems and emphasized the need for measurement and assessment endpoints that are explicitly biological. Various descriptions of the goals of biological monitoring programs have been produced but most concern identifying the presence or effects of an existing or potential human activity on a biological system. For example, the goal of biological monitoring is described, in the narrow sense, by Stevens (1994) as "tracking a particular environmental entity through time, observing its condition, and the change of its condition in response to a well defined stimulus." Kingsford and Battershill (1998) describe an ideal monitoring program as involving "sampling in time with adequate replication to detect variation over a temporal range from short and long time periods, done at more than one location." As pointed out by Kingsford et al. (1998), monitoring studies require repeated sampling through time and should be designed so that the sampling program is able to detect predetermined levels of change in monitored parameters.

Field biological monitoring programs usually have concentrated on population-based abundance parameters such as population density as the biological signal to be monitored (Green 1979; Underwood 1991, 1994). Often only a few key species or "indicator organisms," which have characteristics that make them suitable for detecting impacts (Jones and Kaly 1996), have been targeted for biological monitoring given funding limitations and the logistic inability to study all biological components of any system. However, multimetric indexes (e.g., diversity indexes, index of biological integrity), which require data on multiple taxa and biological conditions, also are often recommended and commonly used to monitor the status of biological systems (Karr and Chu 1997; see chapter 3).

Despite different views on the goals of biological monitoring programs and approaches in their execution, several common requirements for effective field monitoring programs appear to emerge: (1) the biological monitoring program must be carried out over long periods and designed so as to account for natural variability in the biological system; (2) the program must be designed using the best available ecological concepts, study designs, and principles; (3) the data must be collected in a consistent and well-documented manner to achieve required continuity and reliability; and (4) the program should be designed so that the detection of change and impacts can be statistically based.

The most formidable challenge for any successful field-monitoring program is to design an approach that can separate effects of human perturbations from those occurring naturally in the biological system. Natural variability is particularly high in rocky intertidal and other marine systems, and most unsuccessful monitoring programs are

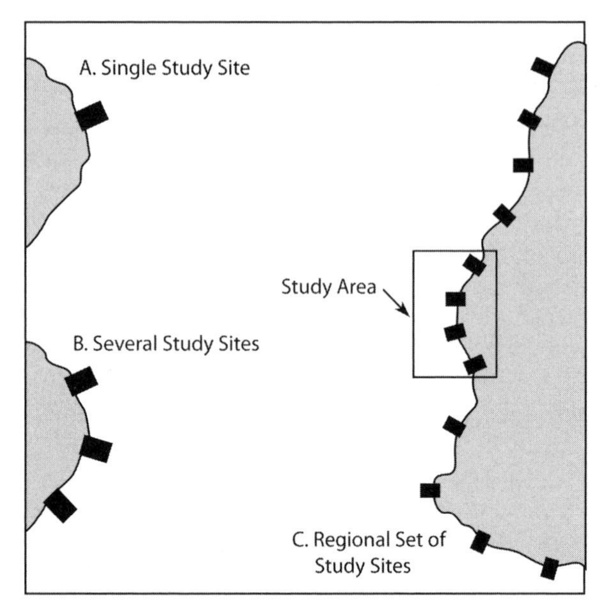

Figure 1.3. Overview of monitoring programs. (A) Program involves only a single study site in the area of interest. (B) Program involves several study sites in the area of interest. (C) Program is part of a regional monitoring program. Only the third approach allows changes at study sites in the area of interest to be placed in the context of regional changes in rocky intertidal populations and communities.

plagued by the inability to separate all but the very largest impacts from the subtler and constantly ongoing changes that take place due to natural processes. In order to understand natural variability in complex rocky intertidal systems, a regional approach is clearly required. For example (as illustrated in figs. 1.3A, B), a single agency or entity interested in monitoring the biological condition of a small section of rocky coastline cannot separate changes due to natural regional processes, such as fluctuations in oceanographic climate, from those caused by local human activities. A much broader regional approach (fig. 1.3C) is required to meet the need for data from multiple sites Kingsford and Battershill (1998). One such approach is that employed in a California rocky intertidal monitoring program being carried out under the leadership of the Minerals Management Service (Ambrose et al. 1995; Raimondi et al. 1999). Funding for broad regional approaches, however, is difficult to maintain over the time scales required to obtain understanding of natural cycles in marine biological systems and almost always must involve coordinated efforts by multiple parties and agencies. Southward (1995) argues that, to be effective, program lengths need to cover the life spans of the dominant organisms in the system and the cycling periods of important environmental factors, which amounted to periods of 11 to 45 years in the English Channel. Timely and coherent ecological information collected over regional and national scales is an integral part of the Environmental Monitoring and Assessment Program (EMAP) initiated by the U.S. Environmental Protection Agency (Stevens 1994).

LITERATURE CITED

Ambrose, R.F., J.M. Engle, P.T. Raimondi, M. Wilson, and J.A. Altstatt. 1995. Rocky intertidal and subtidal resources: Santa Barbara County mainland. Report to the Minerals Management Service. Pacific OCS Region. OCS Study MMS 95–0067.

Carefoot, T. 1977. *Pacific seashores. A guide to intertidal ecology*. Seattle: Univ. of Washington Press.

Ellis, J.I., and D.C. Schneider. 1997. Evaluation of a gradient sampling design for environmental impact assessment. *Environ. Monit. Assess.* 48:157–72.

Green, R.H. 1979. *Sampling design and statistical methods for environmental biologists*. New York: John Wiley and Sons.

Hawkins, S.J., and R.G. Hartnoll. 1983. Changes in a rocky shore community: an evaluation of monitoring. *Mar. Environ. Res.* 9:131–81.

Hurlbert, S.J. 1984. Pseudoreplication and the design of ecological field experiments. *Ecol. Monogr.* 54:187–211.

Jones, G.P., and U.L. Kaly. 1996. Criteria for selecting marine organisms in biomonitoring studies. *In Detecting ecological impacts: concepts and applications in coastal habitats*, ed. R.J. Schmitt, and C.W. Osenberg, 29–48. New York: Academic Press.

Karr, J.R., and E.W. Chu. 1997. Biological monitoring: essential foundation for ecological risk assessment. *Hum. Ecol. Risk Assess.* 3:993–1004.

Kingsford, M.J. 1998. Analytical aspects of sampling design. *In Studying temperate marine environments. A handbook for ecologists*, ed. M.J. Kingsford and C.N. Battershill, 49–83. Christchurch, New Zealand: Canterbury University Press.

Kingsford, M.J., and C.N. Battershill 1998. Procedures for establishing a study. *In Studying temperate marine environments. A handbook for ecologists*, ed. M.J. Kingsford and C.N. Battershill, 29–48. Christchurch, New Zealand: Canterbury University Press.

Kingsford, M.J., C.N. Battershill, and K. Walls. 1998. Introduction to ecological assessments. *In Studying temperate marine environments. A handbook for ecologists*, ed. M.J. Kingsford and C.N. Battershill, 17–28. Christchurch, New Zealand: Canterbury University Press.

Knox, G.A. 2000. *The ecology of seashores*. CRC Press, Boca Raton, Florida, USA.

Kozloff, E.N. 1983. *Seashore life of the northern Pacific coast. An illustrated guide to northern California, Oregon, Washington, and British Columbia*. Seattle: Univ. of Washington Press.

Lewis, J.R. 1964. *Ecology of rocky shores*. London: English Universities Press.

Maher, W.A., P.W. Cullen, and R.H. Norris. 1994. Framework for designing sampling programs. *Environ. Monit. Assess.* 30:139–62.

McKone, M.J., and C.M. Lively. 1993. Statistical analysis of experiments conducted at multiple sites. *Oikos* 67:184–86.

Menge, B.A., and G. Branch 2000. Rocky intertidal communities. In *Marine community ecology*, ed. M.D. Bertness, S.D. Gaines, and M.E. Hay, 221–51. Sunderland, MA: Sinauer Associates.

Osenberg, C.W., and R.J. Schmitt. 1996. Detecting ecological impacts caused by human activities. In *Detecting ecological impacts: concepts and applications in coastal habitats*, ed. R.J. Schmitt, and C.W. Osenberg, 3–16. New York: Academic Press.

Osenberg, C.W., R.J. Schmitt, S.J. Holbrook, K. Abu-Saba, and A.R. Flegal. 1996. Detection of environmental impacts: natural variability, effect size, and power analysis. In *Detecting ecological impacts: concepts and applications in coastal habitats*, ed. R.J. Schmitt, and C.W. Osenberg, 83–107. New York: Academic Press.

Paine, R.T., J.L. Ruesink, A. Sun, E.L. Soulanilee, M.J. Wonham, C.D.G. Harley, D.R. Brumbaugh, and D.L. Secord. 1996. Trouble on oiled waters: lessons from the *Exxon Valdez* oil spill. *Annu. Rev. Ecol. Syst.* 27:197–235.

Raffaelli, D., and S. Hawkins. 1996. *Intertidal ecology*. London: Chapman and Hall.

Raimondi, P.T. R.F. Ambrose, J.M. Engle, S.N. Murray, and M. Wilson. 1999. Monitoring of rocky intertidal resources along the central and southern California mainland. 3-year report for San Luis Obispo, Santa Barbara, and Orange Counties (Fall 1995–Spring 1998). Technical Report MMS 99–0032. U.S. Department of Interior, Minerals Management Service, Pacific OCS Region.

Ricketts, E.F., J. Calvin, J.W. Hedgpeth, and D.W. Phillips. 1985. *Between Pacific tides*. 5th ed. Stanford, CA: Stanford University Press.

Southward, A.J. 1995. The importance of long time-series in understanding the variability of natural systems. *Helgoländer Meeresunters* 49:329–33.

Stephenson, T.A., and A. Stephenson. 1972. *Life between tidemarks on rocky shores*. San Francisco: W. H. Freeman.

Stewart-Oaten, A. 1996. Problems in the analysis of environmental monitoring data. In *Decting ecological impacts: concepts and applications in coastal habitats*, ed. R.J. Schmitt and C.W. Osenberg, 109–31. New York: Academic Press.

Stewart-Oaten, A., W.W. Murdoch, and K.R. Parker. 1986. Environmental impact assessment: "Pseudoreplication" in time? *Ecology* 67:929–40.

Stevens, D.L., Jr., 1994. Implementation of national monitoring program. *J. Environ. Monit.* 42:1–29.

Suter, G.W., ed. 1993. *Ecological risk assessment*. Boca Raton, FL: Lewis.

Underwood, A.J. 1991. Beyond BACI: experimental designs for detecting human environmental impacts on temporal variations in natural populations. *Aust. J. Mar. Freshwater. Res.* 42:569–87.

———. 1992. Beyond BACI: the detection of environmental impacts on populations in the real, but variable world. *J. Exp. Mar. Biol. Ecol.* 161:145–78.
———. 1993a. Field experiments in intertidal ecology. In *Proc 2nd Int Temp Reef Symp*, 7–13.
———. 1993b. The mechanics of spatially replicated sampling programmes to detect environmental impacts in a variable world. *Aust. J. Ecol.* 18:99–116.
———. 1994. On beyond BACI: sampling designs that might reliably detect environmental disturbances. *Ecol. Appl.* 4:3–15.
———. 1997. *Experiments in ecology: their logical design and interpretation using analysis of variance*. Cambridge: Cambridge Univ. Press.

Zonation of rocky shore organisms at an exposed site in Olympic National Park, Washington State.

CHAPTER 2

Site Classification and Selection

For any intertidal sampling program, site selection probably has the greatest influence on the overall program design. As discussed in later chapters, the answer to "Where to sample?" depends on the program goals. Is the goal to characterize the health of an entire coastline or to determine the impact of pollutants at a specific site? Do we need to infer from our data how a population is changing 100 km up the coast or simply how many species were at a site before oil hit it? Are we trying to detect impacts on all intertidal substratum types or only on flat rocky benches? Unless sampling is intentionally confined to one location, and site comparisons are not needed, site selection is an important decision—one on which all conclusions obtained from the sampling program will be based. If sites to be compared (e.g., control sites vs. impact sites or multiple sites along a gradient of anthropogenic influence) are not physically similar, then physical differences can confound any conclusions about what might cause or correlate with any detected differences or trends.

This chapter discusses procedures for site selection (fig. 2.1), including methods of classifying shorelines to avoid confounding problems resulting from differences in geophysical site characteristics. The utility of shoreline habitat maps and the pros and cons of sampling particular microhabitats on rocky shores also are reviewed. Additionally, how sites should be selected based on program goals and how these goals will differ depending on the objectives of the sampling program are discussed. Finally, the key role of replicating sites in developing robust study designs is addressed.

CLASSIFYING SHORELINES

Intertidal ecologists have long noted that factors such as wave exposure, amount of shade, and even rock type can affect species distributions and abundances. These physical features provide the backdrop against which

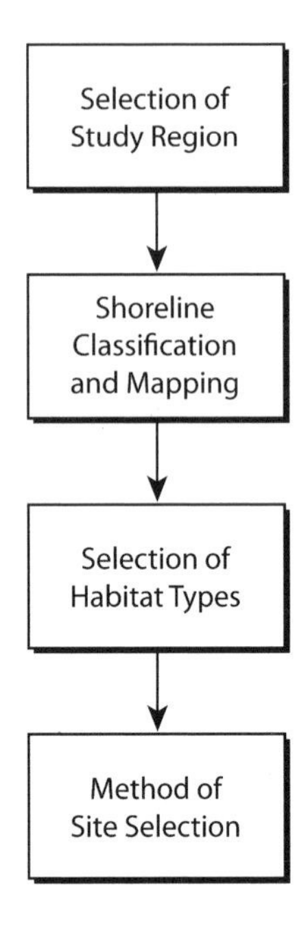

Figure 2.1. A recommended approach for site selection for comparative studies of control and impacted study areas.

biological interactions occur, and the presence or absence of a particular species in a community depends on having an appropriate physical environment. To select ecologically comparable sites (and thus avoid confounding influences during data analyses), we must control for these variables as thoroughly as possible. For example, Schoch and Dethier (1996) found that within a 5-km stretch of Washington shoreline, there were significant differences in community composition between rocky sites varying only in relatively subtle physical characteristics, especially slope. Site selection thus must begin by classifying potential study sites according to features important to the structure and dynamics of rocky shore communities.

Geophysical Features

When are sites similar enough to allow valid comparisons between their biota? No one would compare a sand beach with a mud flat, but all rocky shores are not the same and comparing sites with different wave exposures,

TABLE 2.1. Geophysical Variables to Consider When Choosing Sites for Monitoring or Impact Assessment Study Programs

Variable	Rationale: Parameters Affected
Wave Exposure	Delivery of food, nutrients, gases, and propagules; ease of attachment; wetting of surfaces; foraging capability of some predators
Nearshore Water Depth	Wave energy delivered to the site
Aspect	Desiccation; heating; exposure to waves and currents
Slope and Width	Insolation; wave run-up; sometimes ease of settlement and attachment for organisms; foraging capability of some predators
Roughness	Provision of habitats for predators, cryptic species, desiccation-sensitive species, and wave-sensitive species
Rock Type	Heating; small-scale surface roughness; water retention
Degree of Sand Influence	Cover of opportunists; presence of sand-loving and sand-resistant species
Degree of Ice Scour	Amount of bare space; presence of resistant or opportunistic species
Salinity	Presence and absence of many taxa in estuaries; increased mortality of species intolerant of salinity change in pools and crevices
Human Usage	Reductions in the abundances of species sensitive to trampling and collecting activity; changes in population structure of extracted species toward higher frequencies of smaller-sized individuals

NOTE: For details see Schoch and Dethier (1996).

rock types, proximities to a river, or aspects can be very misleading. Table 2.1 offers a minimal checklist of physical factors important to rocky shore community structures. Screening for these factors during the early stages of study design should be part of any sound site-selection procedure. Besides the actual substratum type (e.g., rock vs. gravel), the most important attribute affecting intertidal community structure at landscape-level scales is probably wave energy. The other factors listed in table 2.1 are probably of secondary importance, although research elucidating their relative roles in influencing community structure is lacking. At larger spatial scales, factors such as tidal range and regional climate probably also lead to the differentiation of intertidal communities. At smaller-than-landscape

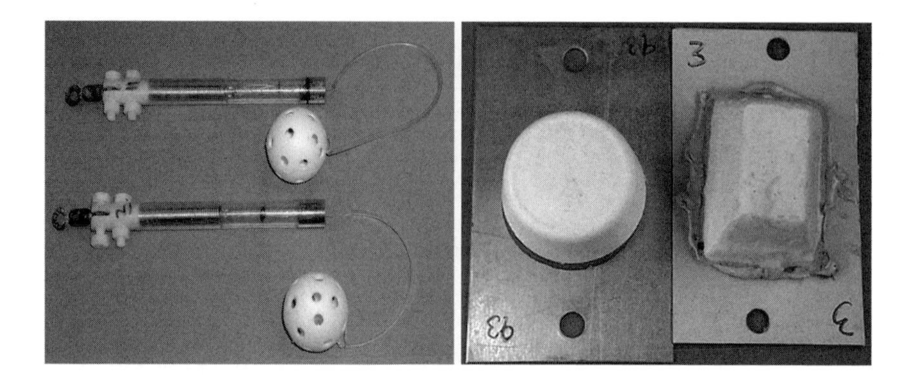

Figure 2.2. (A). Wave force meters as designed by Bell and Denny (1994). (B) Dissolution modules or clod cards similar to the model designed by Doty (1971).

scales—for example, where substrata and wave energy are fairly homogeneous (e.g., exposed rocky shores)—other factors such as surface roughness and human disturbance gain importance. Of course, the list in table 2.1 is not exhaustive; for instance, Menge et al. (1997) recently demonstrated that the degree of nearshore upwelling may have significant background effects on intertidal community structure, such that otherwise-similar rocky shores develop different patterns of species composition and abundance. In addition, any comparison of biota among sites or among times must carefully control for tidal height, since emersion time strongly affects the biota. Actual (as opposed to predicted) emersion time will be affected by wave run-up as discussed below. When comparing sites with different tidal ranges (as seen e.g., in NOAA tide tables), it may be useful to sample each site at a local datum (e.g., mean lower low water or mean sea level) to compensate for these differences.

Wave Energy. Wave energy is manifested in various ways. For example, wave velocity can tear organisms off rocks or can roll the substratum if particles are of critical rolling diameter, whereas wave run-up affects the vertical elevation of the community by reducing desiccation stress at higher tidal heights. Measuring wave exposure is one of the most difficult yet most important tasks for classifying rocky shores.

There are three general procedures for measuring and expressing the degree of wave exposure experienced by a rocky shore: (1) estimating cumulative water motion or maximum wave force directly, using some kind of gauge, meter, or dissolution module (fig. 2.2) (e.g., Denny 1985, Bell and Denny 1994, Raffaelli and Hawkins 1996, Castilla et al. 1998);

TABLE 2.2. Wave Parameters Derived for Calculating the Surf Similarity Index.

Energy Category	Exposure Classification	Fetch Distance (km)	Sustained Wind Speed (kts)	Significant Wave Height (m)	Wave Period (sec)	Wave Length (m)
1	Very protected	<0.1	5	0.1	1.0	1
2		0.1–0.5	10	0.2	1.5	2
3		0.5–1	10	0.3	2.0	6
4	Protected	1–5	20	0.4	2.5	10
5		5–10	20	0.5	3.0	14
6	Semiprotected	10–50	30	1.0	4.0	25
7	Semiexposed	50–100	30	2.0	5.0	40
8		100–500	40	3.0	6.0	60
9	Exposed	500–1000	40	4.0	8.0	100
10		>1000	50	5.0	10.0	150

SOURCES: Exposure classification from Harper et al. (1991); other data from CERC (1984). Reprinted from Schoch and Dethier (1997).

(2) calculating wave energy based on other measured parameters (e.g., Denny (1985 and table 2.2); and (3) estimating wave energy indirectly, using biological or physical indicators. Wave exposure also is often reported based on qualitative estimates that may be assigned to a numerical scale (e.g., 1 for extremely strong wave exposure and 5 for essentially no wave exposure). Direct measurements should be taken over a long time period (e.g., multiseasonal) and under a range of conditions to best integrate the wave energies experienced by a site. Sites being matched for comparative study should have their wave energies quantified simultaneously to control for temporal variation in measured parameters.

Indirect measurements include biotic indicators, such as the presence of indicator species, the width of the intertidal zone, and the height of the splash or other distinct zones above low water, and physical indicators, such as the fetch or distances traveled by waves prior to reaching the site. Maximum fetch, for example, is often used in such indirect estimates of wave exposure because this is the longest distance over which waves travel to the site unimpeded by a landmass. Wave statistics also can be estimated using graphs published in the *Shore Protection Manual* (Coastal Engineering Research Center [CERC] 1984). Table 2.2 provides an example of this indirect approach using the graphs for a large coastline with a broad range of energy levels. Harley and Helmuth (2003) describe a novel approach using temperature dataloggers to estimate the key effect of wave splash on zonation patterns. If investigators are interested

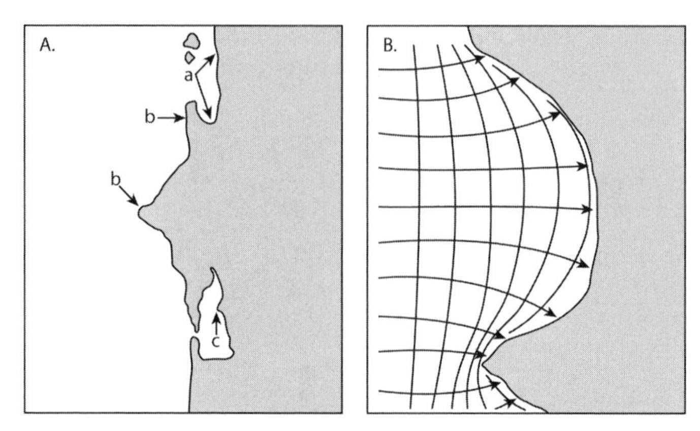

Figure 2.3. (A) Types of coastline and (B) patterns of wave refraction showing convergence of wave energy on rocky headlands. Site *a*, protected outer coast; sites *b*, exposed headlands; site *c*, embayment. (Redrawn from Ricketts et al. 1985.)

only in relatively quiet bays or summertime (nonstormy) conditions, then categories encompassing a narrow range of low wave energies could be used to form such site groups.

Nearshore water depth, which affects the wave energy delivered to a rocky shore, also can be used to indirectly estimate wave exposure. Depth profiles usually can be adequately determined from detailed marine charts for the region near a study site. However, the scale over which the chart provides depth information must match the project's scale. Sites where the ocean bottom drops off steeply near the shore, or where a deep submarine canyon approaches the coast, will receive greater wave forces than sites with long, shallow seaward approaches. Nearshore topography also can focus or diffuse waves; waves will refract and focus their energy on headlands, for example, while embayments will receive more diffuse and reduced wave energy (fig. 2.3).

Aspect and Slope. The aspect, or primary compass direction of a site, is important in terms of both wave and sun exposure. If a beach is oriented in the direction facing prevailing swell patterns or large seasonal storms, considerable disturbance can be expected for communities occupying that face. Similar levels of disturbance, however, may not characterize communities facing other directions. In the Northern Hemisphere, shores facing to the south receive more direct sunlight and experience different desiccation and temperature extremes than north-facing shores. Site aspect can be determined readily from a detailed map or from on-the-ground surveys. Areas that comprise a rocky headland (fig. 2.3) present more complexity compared with longshore habitats and should probably be considered at least two sites with different aspects—for example, a north-facing site as well as a south-facing site—because of their multiple faces with respect to sea and sun conditions.

Slope is important because of its effects both on the dissipation of wave energy and on insolation patterns. Steep shores will reflect waves. Low shore angles will dissipate wave energy, and intermediate angles will be the most disturbed by wave action because the wave energy will be expended suddenly on the beach face. Thus, shore slope may have a greater effect on the wave energy organisms experience than the run-up or maximum fetch over which waves travel before breaking on the shore. A high-shore area bordered by a long, flat, midshore bench will effectively experience less wave exposure than one with a steep mid- and low-shore region. Slope also may affect whether or not spilled oil will strand on a shore and, if oil deposition occurs, influence its pattern of distribution. The slope of a shore can be measured directly with an inclinometer or estimated from knowledge of tidal amplitude and the width of the intertidal zone; the latter information can be obtained from aerial photographs or videos taken at low tide.

Topographical Heterogeneity and Rock Type. Topographical heterogeneity or roughness of rock surfaces creates turbulence that can dissipate wave energy (thus providing sheltered microhabitats) and possibly enhance nutrient/food and waste exchange. Topographical heterogeneity can be important on several spatial scales and influence species diversity and abundance (Archambault and Bourget 1996). Large (site)-scale roughness, for example, the relative abundance of topographic features such as crevices, pools, and hummocks, affects local distributional patterns by providing a variety of microhabitats for sessile and mobile organisms. This feature is perhaps best quantified on a subjective scale, for example, a score of 1 for a site comprised of relatively unbroken, even bedrock, and 5 for a site where numerous features make the surface very irregular on a scale of meters. On a less coarse scale, variations in substratum topography can be measured using a transect tape and chain (Luckhurst and Luckhurst 1978). To apply this method, the chain is laid out along the bottom contours while the transect tape is pulled taut to measure the linear distance between the start and the finishing points. The ratio of the surface distance (determined by the chain) to the linear distance (determined by the transect tape) is used as a measure of substratum rugosity.

Surface roughness also can be described on a much smaller scale, where it relates closely to the type of rock, for example, the fine crystalline texture of a rocky surface or the abundance of tiny cracks and

Figure 2.4. A form of profile gauge for measuring topographic heterogeneity. An array of regularly spaced rods is dropped perpendicularly from the leveled platform and the distance to the substratum is measured. These data can be used to calculate a topographic index using procedures described by Underwood and Chapman (1989).

pores in vesicular volcanic rock. The importance of this textural roughness is not well known, but roughness could increase water retention and provide refugia for newly settled benthic recruits (Littler 1980) or could serve to hold contaminants such as oil on the shore. Often surface roughness is quantified using a profile gauge, where the linear distances that randomly or regularly arranged rods are dropped perpendicular to the substratum from a level, horizontal frame or Plexiglas (Perspex) board (fig. 2.4) are used to calculate a topographic index reflecting substratum microrelief (Underwood and Chapman 1989). Unfortunately, little research has been done on the effects of physical and chemical differences among rock types on rocky shore communities, although there is evidence that some organisms recruit or survive differentially on different substratum types (e.g., Stephenson 1961; Raimondi 1988; Lohse 1993). The tendency of some rock types to readily erode (e.g., sandstone) or fracture (e.g., some metamorphics) may also affect the development and persistence of sessile organisms on rocky shores.

Sand Burial and Scour. The frequency and magnitude of sand burial or scour is an important variable influencing the structure and organization of rocky shore communities. For example, sand affects the distribution and abundances of rocky intertidal populations along much of the southern California mainland coastline (Murray and Bray 1993), where the upper parts of shores are frequently covered and uncovered by sand (Littler et al. 1991). Sand-scoured rocky sites differ from those lacking sand influence, both in species composition and in the abundances and dynamics of resident populations. Three categories of macrophytes described by Littler et al. (1983) are found in abundance on southern California rocky shores that receive high levels of sand influence (Murray and Bray 1993): (1) opportunistic seaweeds that rapidly colonize disturbed substrata, (2) resistant macrophytes that tolerate sand abrasion and burial, and (3) "sand-loving" or psammophytic algae, which, for unknown reasons, are best represented in sand-influenced communities.

As with wave exposure, quantification of sand scour and sand burial is very difficult to perform. Researchers have used sediment traps (Airoldi and Cinelli 1997; Airoldi and Virgilio 1998) and abrasion rates of concrete blocks (Craik 1980) to measure sand influence. More common approaches are to make repeated measurements of the depth of sand at site reference stations or to rank sites in terms of sand exposure based on qualitative observations (e.g., 0 indicating no sand and 10 indicating frequent burial under thick sand). However, none of these approaches seems to fully reflect the role played by sand movements on rocky shore populations. Instead of sand, the degree or frequency of ice or driftwood scour or longshore ice flow may be important on some rocky shores and play a role in study-site selection (e.g., Dayton 1971; Wethey 1985; Kiirikki 1996; Pugh and Davenport 1997; McCook and Chapman 1997).

Salinity. Salinity generally varies little along the open coast and is not considered to be an important factor in structuring rocky intertidal communities; however, salinity can be a very important factor in determining species distributions and abundances in estuarine situations or near the mouths of coastal rivers (reviewed in Ardisson and Bourget 1997; Witman and Grange 1998). Sites near river mouths, or even openings of smaller streams, can be affected not only by variations in salinity but also by the sediment load and contaminants brought down to the shore from coastal watersheds. In addition, evaporation and rainfall can cause salinity to vary considerably on high-shore rock or pools, especially in wave-protected areas. Salinity can be measured directly at sites using a refractometer or portable conductivity meter; both types of devices are subject to calibration errors and should be checked for accuracy against known standards. As with wave action and other features of the physical environment that can vary considerably over short temporal scales, where salinity may be an important environmental factor measurements should be performed at study sites at approximately the same time and the same depth to optimize site comparisons.

Human Disturbance. Finally, human disturbance (e.g., trampling, collecting, and tidepool exploring) may seriously alter the biota of heavily used rocky shores (Murray et al. 1999), such that even physically matched sites will have different community structures and dynamics if levels of human disturbance differ. Sites subjected to high levels of visitor foot traffic (Brosnan and Crumrine 1994) and exploratory activities (Addessi 1995) may show reduced abundances of vulnerable species and increased abundances of disturbance-resistant and opportunistic taxa. Collection of organisms for human consumption, fish bait, or other purposes can cause exploited species to exhibit reduced densities and altered population size structures at heavily impacted sites (Griffiths and Branch 1997). There is an increasing effort in marine parks to gather data on levels of human impact. If available, such data could be used to create a ranking of this variable when popular sites are needed for a monitoring or impact-detection program.

Mapping Methods

If all shorelines worldwide were mapped with detailed information on the geophysical characteristics just described, it would be relatively simple to begin a study by identifying a pool of sites that meet targeted criteria. Because such maps do not exist for most areas, however, the alternatives are either to map the section of coastline under consideration, then identify potential study sites, or to choose a range of potential sites in the region, then screen them out according to selected criteria. Choosing between these options is likely to depend on funding and the time available before the study is to be initiated because it takes time and fiscal resources to develop useful maps.

Mapping can be done at a wide variety of scales, depending on the level of resolution needed for the study, the size and complexity of the shorelines being mapped, and the available funds and equipment. Maps of an entire coastline designed to display the overall abundance of rock and sand, for instance, could probably be prepared using coarse-scale data such as aerial photographs (e.g., with color infrared photography) or satellite images. A central repository of maps, aerial photographs, videos, and other information providing data on shoreline geophysical characteristics would be of substantial value for first-cut site-selection decisions. This would be particularly true in the event of an oil spill because it is unlikely that time would be available to gather and analyze the geophysical information needed to set up intertidal study sites during the short period that spilled oil is at sea. Unfortunately, central repositories generally do not exist and investigators must usually gather their own geophysical data from a variety of sources in preparing for study site selection.

Information on the physical features of shorelines has sometimes been taken from environmental sensitivity maps (e.g., Research Planning, Inc. 1990), but these are generally out-of-date, sometimes inaccurate, and often not of high enough resolution to inform site selection. Such qualitative maps, if used as the sole information source for selecting similar sites, are likely to lead to later problems in data analysis and interpretation because the degree of variability inherent in some of the categories (e.g., wave-cut platforms and exposed pier structures) means that the biota are likely to be similarly variable. If the geophysical and biotic characteristics of the sites selected for study vary greatly, then the ability to detect spatial or temporal differences in species distributions and abundances will be greatly weakened by the loss of statistical power that comes with high variance (see chapter 4).

Helicopter overflights supplemented with photographs and videos offer finer-resolution data (e.g., the ability to distinguish different types of unconsolidated beaches). This method was used to map the geophysical features of much of the Washington and British Columbia shoreline (Harper et al. 1991; Howes et al. 1994) and the coarse-scale physical features and biotic assemblages of intertidal habitats (fig. 2.5) throughout southern California (Littler and Littler 1987). The finest-resolution data come from trained observers boating closely or walking along the shoreline, gathering quantitative information on the physical characteristics described earlier (Schoch 1996), or from periodic helicopter landings for ground-truthing site characteristics viewed from the air (Littler and Littler 1987). If physical factors during such surveys are measured quantitatively (i.e., using categorical or continuous data), it is easier to avoid the error of assuming that sites are similar when they may have an accumulation of small but significant differences. If data are entered eventually into a digital mapping format such as a Geographical Information System (GIS), then many attributes can be included and used for site screening. Practically speaking, sites smaller than 10 or even 100 m in the longshore direction will be too small to be resolved when viewing a regional map. A useful approach for informing site selection is to map at scales greater than 10 m, seeking to compare beach segments that are physically homogeneous at this level of resolution.

One system for gathering quantitative geophysical information is the SCALE (Shoreline Classification and Landscape Extrapolation) model (fig. 2.6), which is designed to define shorelines precisely enough that the

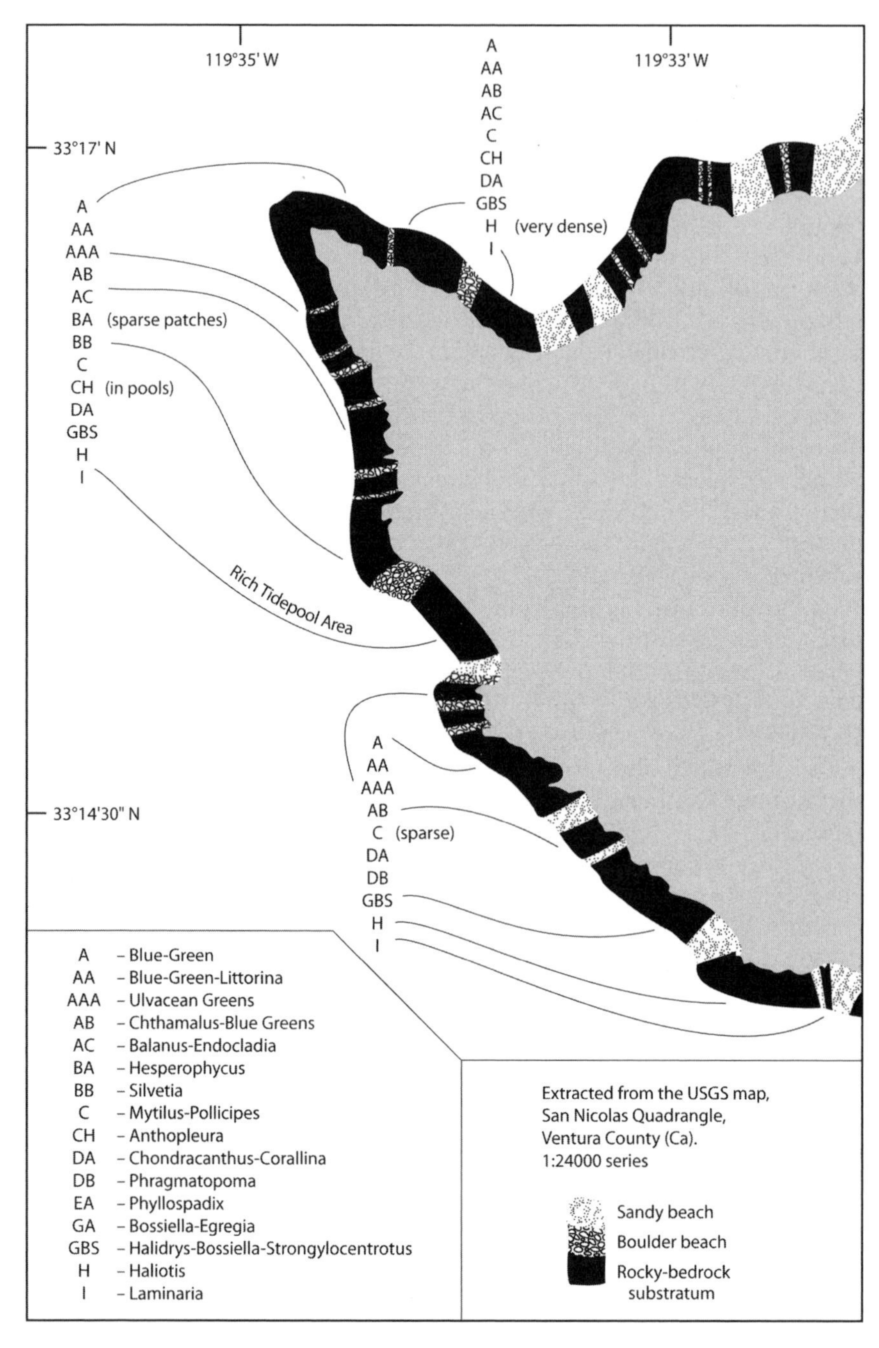

Figure 2.5. Example of a section of the coastline on San Nicolas Island in southern California categorized by Littler and Littler (1987) using helicopter overflights. (Redrawn from original figure supplied by Mary Elaine Dunaway, Minerals Management Service (Pacific Region), U.S. Department of the Interior.)

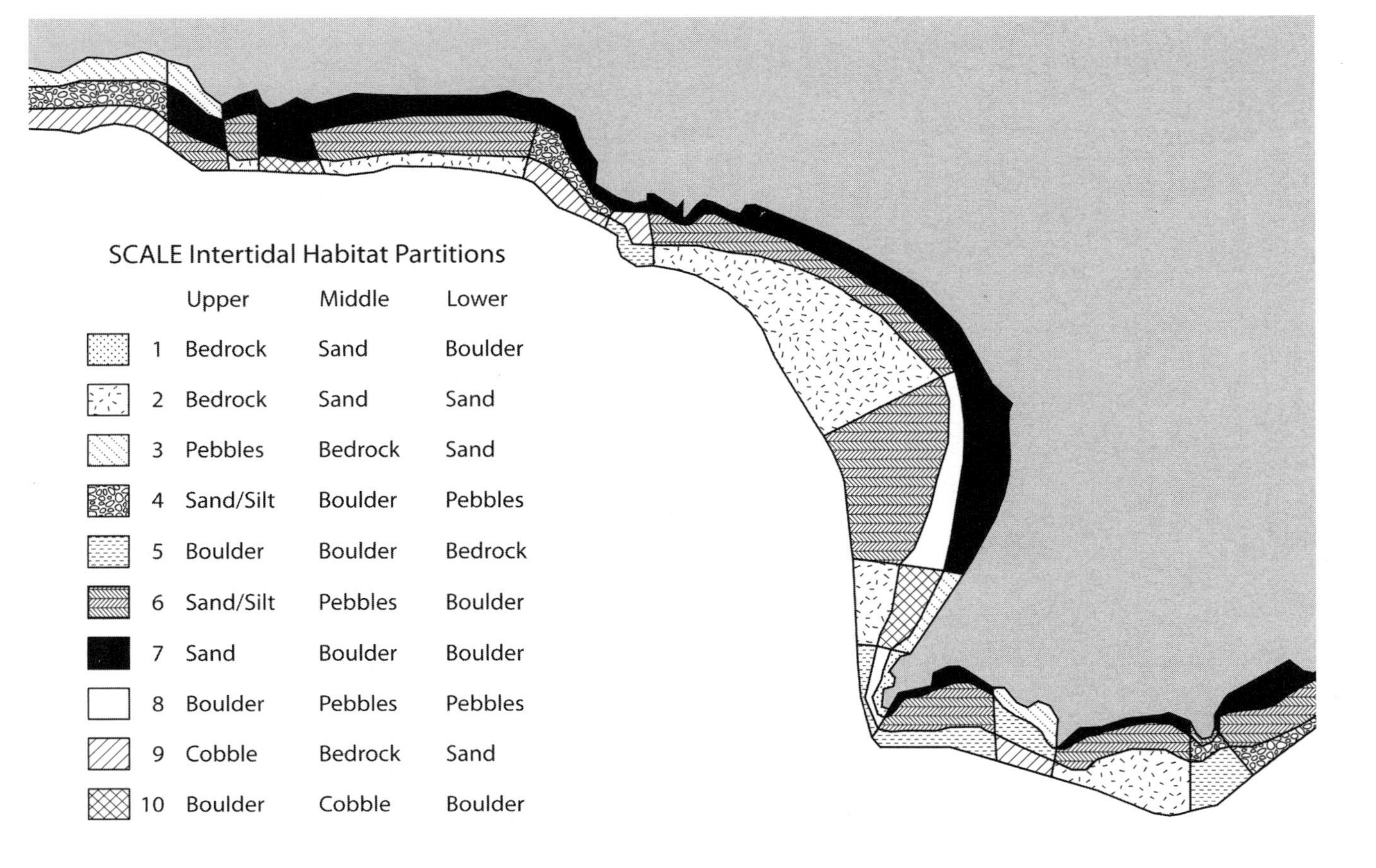

Figure 2.6. SCALE map of a section of a Pacific Northwest shoreline showing key geophysical features of nearshore habitats.

biota on statistically similar shores should be very similar (Schoch 1996; Schoch and Dethier 1997). This model was designed for shorelines characterized by a broad range of wave energies and substratum types and for addressing questions involving change detection at the scale of species populations and sites. If shorelines are classified into geophysically homogeneous beaches, then each beach should in theory have relatively homogeneous biota presuming similar disturbance histories (Schoch and Dethier 1996). Such homogeneous beach groupings can serve as high-resolution mapping units for informing study site selection. The concept of using geophysical characteristics to predict potential biotic communities should hold for any coastal region. Of course, ecological history (e.g., exposure to disturbance events, past recruitment success) and biotic interactions (e.g., level of predation) will affect the current status of rocky intertidal populations and communities, resulting in biotic variation among sites that cannot be predicted by models built solely on geophysical site features. We can never entirely remove these sources of variability, but choosing geographically close and physically similar sites will reduce their influence on collected data.

HABITAT TYPES

Once shoreline classification has been done, the specific habitat types to be sampled must be determined. While this handbook focuses on rocky shores, even within this broad habitat type there are numerous decisions to be made about the specific features of the habitats selected for study. This is because rocky intertidal habitats are highly variable and can be differentiated based on wave exposure, slope (e.g., steep versus flat benches), topographical heterogeneity, substratum composition and nature (e.g., bedrock versus boulder fields), and even the presence and frequency of microhabitats such as tidepools and crevices (fig. 2.7). For each of these variable habitat types, sampling can be performed either across all tidal elevations from the high tideline to the lower limits of the intertidal zone or only at selected tidal heights. On the other hand, sampling may be focused on biotically generated habitats such as surfgrass (*Phyllospadix* spp.) or mussel (*Mytilus* spp.) beds.

Given that rocky shores come in many forms, how should investigators target habitats for study? Due to logistical constraints, it will almost always be impossible to sample all habitats effectively. If the goal of the study is to select habitat types for a comprehensive monitoring program, then key questions might be what shoreline/habitat types are of greatest concern or interest and what types are most at risk. Habitats of greatest concern might be those with the highest biodiversity, or those able to support populations of a particular species (such as abalone), or simply those most common

Figure 2.7. Examples of various rocky intertidal habitat types. (A) Strong wave exposure at Botany Beach, Vancouver Island, British Columbia. (B) Topographically heterogeneous habitat at Torch Bay, Alaska. (C) Broken bedrock and topographically heterogeneous habitat at Dana Point, southern California. (D) Sand-influenced rocky intertidal habitat at Crystal Cove, southern California.

along a coastline; this decision is value-based. Habitats most at risk might be flat rocky benches, because they readily collect oil and often receive the greatest intensity of human visitation, or protected shores, where pollutants may be more persistent. This decision must be based on the goals of the study program, for example, the likelihood of site exposure to current or future environmental stressors. Impact detection programs may justifiably focus on a constrained subset of habitat types. For example, if oil from a spill comes ashore only in the high zone, there will be limited utility in sampling biota occupying the diverse and complex low intertidal zone.

Researchers may choose to target unusual habitat types because of their special value or vulnerability. Boulder fields, for example, contain a variety of microhabitats including the undersides of rocks and the sediment beneath them, as well as the more obvious visible (tops and sides) rock surfaces. Given that this habitat type is common in some coastal regions, is used by humans, is biologically diverse, and is likely to retain oil following a spill, it is probably an important one to sample for both monitoring and oil-impact assessment purposes. Tidepools often are a focus of public interest given that they have aesthetic appeal, may contain subtidal species that only rarely occur in intertidal habitats, and tend to be very accessible. However, because each tidepool is physically and biotically unique, information gained from studying one pool is unlikely to apply (at least quantitatively) to other pools. This uniqueness makes tidepools less useful as a focus of a comparative sampling program. Marking individual pools and quantifying their biota through time may provide information about changes that occur in that particular pool (e.g., from being visited by too many school children), but this approach cannot be used to detect broad regional trends in pool biota. However, studies of the impact of oil, for example, in a few pools at one site, may provide some indication of possible impacts in nearby pools and be useful for impact assessment.

Structure-forming biota, such as mussels, surfgrass, and some tubeworms, may be a worthwhile focus of sampling because these bioengineered habitats are diverse and particularly vulnerable to some kinds of impact. For example, if an oil spill or other form of disturbance kills or reduces the abundance of the key structure-forming species, the entire habitat with its associated biota will be impacted. In addition, because these biotically structured beds often are long lived, their destruction is of particular concern. Chapters 4 and 5 discuss unique problems involved in sampling some of these habitat types.

METHODS FOR SITE SELECTION

Regardless of the program goals, it is important to consider geophysical features if a study is being designed to compare populations or communities among sites. Beyond this general caveat, methods of site selection will differ depending on the question being asked. In this section, we discuss site-selection considerations for designing resource monitoring and impact determination studies.

Monitoring Studies

The ultimate goal of site selection for resource monitoring is to obtain data of long duration and low noise-to-signal ratio (i.e., low temporal and spatial variability relative to the effect or trend being quantified), so that we can detect biological patterns that can be analyzed and interpreted. A recent report from a workshop on ecological resource monitoring (Olsen et al. 1997) distinguished three possible types of monitoring sites that represent different scales of investigation: (1) survey sites (such as EMAP, the Environmental Monitoring and Assessment Program, described below) that are designed to be statistically representative of habitats on regional or national scales and where the gathered information may have management or policy implications; (2) networks of sites that are used for monitoring large (regional)-scale patterns of biological resources and environmental stressors; and (3) intensive (or sentinel) sites, which tend to be few and are used to generate process-level understanding but which cannot be treated like statistical samples or directly used to represent regional trends or patterns in biological parameters.

For each type of monitoring site, site selection can rely on either design-based or model-based inference. Design-based (or probability-based) studies justify inference to unmonitored sites because they rely on statistically random site selection procedures. These require few assumptions, but allow no flexibility in making site selections. For example, the EPA's EMAP uses a sampling design for terrestrial systems involving sampling every 4 years at points on a large, hexagonal grid covering the landscape of the United States, with points (at the landscape characterization scale) spaced 27 km apart. EMAP's primary goal is to "estimate the current status, trends, and changes in selected indicators of condition of the nation's ecological resources on a regional basis." However, there is concern that this sampling program, while statistically valid, is operational "at too coarse of a scale in space and time to detect meaningful changes in the condition of ecological resources" (National Research Council [NRC] 1995, 3). This is an issue of matching sampling design to program goals and figuring out how to allocate personnel and funds to maximize the information gained. The NRC recommended increasing sampling frequency and intensity at a subset of the current sites and using stratified random sampling within regions to achieve particular data-quality objectives.

Model-based procedures, in contrast, use a statistical design to draw conclusions about observed study sites, and any inference to unobserved sites then depends on the statistical correctness of the study design or model. For instance, understanding of causal ecological mechanisms can be gained from data collected at observed sites and then be used to extrapolate the application of these same mechanisms to unobserved sites. The potential problems here are the (unquantifiable) biases introduced with site selection and the assumptions about the broad validity of the model.

The need to develop regional networks or sets of monitoring sites for shorelines, although not at EMAP scales, seems to be increasingly appreciated. Programs that have established regional networks or sets of monitoring sites often have a very broad or comprehensive goal, for example, seeking trends of any kind or changes caused by any processes at some indeterminate time in the future. Such monitoring may have *retrospective* goals—that is, the causes of change or stress are not known beforehand—or *prospective* goals, where stressors are known and thus monitoring can be designed to be more targeted and more efficient. If likely stressors are known, it may be possible to stratify site selection on the basis of the site's responsiveness to a stressor of concern (for example, rocky intertidal sites with habitats that are particularly sensitive to spilled oil). In this way, a *responsive* site might represent a class of sites that should respond in a predictable and similar way to that stressor. However, rarely do we have data that enable generalizations about such site responsiveness.

For retrospective sampling, monitoring is likely to be done at selected sites with the hope that trends and changes can be extrapolated over a larger geographic area. However, having a broad scale of inference requires choosing sites in a statistically and ecologically valid fashion. If sites are not randomly chosen, then inference from data collected at a few sites is limited statistically to those sites, and cannot be used to generalize (see Page et al. [1995] and Beck [1997] for discussions about sampling designs and generalization) or to extrapolate findings to other sites in the region.

An ongoing program at scales smaller than EMAP, but using similar rationales, is the Southern California Bight Pilot Project, carried out by the Southern California Coastal Water Research Project (SCCWRP) (see www.sccwrp.org for information). This project has the goal of defining and monitoring the extent and distribution of environmental degradation over the southern California continental shelf. Site selection here follows a probability-based sampling design. First, the region was divided into areas or subpopulations of interest (e.g., geographic zones, areas around sewage treatment plants), and then sites were chosen randomly within these areas. This design ensures unbiased estimates of the Bight condition and enables determination of the actual span of shelf where ecological conditions differ from reference areas. The scale of inference is thus the whole Bight, although statistical inferences also can be made about each area (e.g., the conditions around river and storm drain outfalls).

The selection of intertidal sites based solely on statistical criteria is likely to result in serious problems if selection is made from inadequate base maps. For example, following the *Exxon Valdez* oil spill, researchers representing the state of Alaska randomly selected study sites based on coarse-scale maps of shorelines showing substratum types. However, when the sampling crews initiated their fieldwork, they found that they were trying to compare sites varying widely in salinity and even, in some cases, in substratum type, thereby significantly reducing the power of their comparative sampling program (McDonald et al. 1995).

Other filters to apply to a population of potential sites before randomly choosing actual sites for study include (1) accessibility and other logistical concerns; (2) size, that is, the length of the segment of relatively homogenous shoreline—this is because short beach segments will suffer from edge effects such as sand scour at the edge of a rocky bench; (3) proximity to known stressors (e.g., sites near and far from an outfall site or sites located on and away from modeled oil spill trajectories); and (4) desired geographic spread or range of the targeted study area. If comparable sites have been selected, then those spread over a broad geographic range are more likely to let us distinguish local events from those of a climatic or hydrographic nature: "The geographic scale of any biological pattern or trend is the first and most reliable clue to its cause" (Lewis 1982). If a common pattern is seen over a large area, then any departure from this pattern is strong evidence for a local impact—from a pollutant, for example (e.g., Bowman 1978; Christie 1985).

A rather different sampling scheme from the network approach involves using "sentinel" or "reference" sites. These are intensively sampled areas, ideally used to provide a linkage between a broad-scale survey (extensive but low-resolution) and localized, process-oriented basic research that enables understanding of cause and effect. Jassby (1998) argues that if sentinel sites can elucidate key mechanisms of interannual variability, then they may be useful for generalizing to other sites sharing their main features. For example, individual lakes have certain characteristics (e.g., annual ice cover, morphometry of the watershed) that predispose them to be responsive to a certain set of external forces (e.g., acid rain). If we can understand these predisposing features for particular stressors on rocky shores, then we may be able to extrapolate from sentinel sites containing these features to other sites that share them. This could be regarded as a special case of a "model-based" design.

Disadvantages of relying on a monitoring strategy that incorporates sentinel sites are that (1) considerable knowledge of the "natural history of interannual variability" is required (Jassby 1998); (2) background variability (among populations, sites, or years) can disguise a system's responsiveness to a certain stressor; and (3) sentinel sites chosen because they are sensitive to change may exaggerate the amount of change occurring in a region. However, intensively monitored sentinel sites, especially if they can be chosen in a random (or stratified random) fashion from a population of potential sites, may provide the best way to understand the mechanisms responsible for long-term biotic change in a region.

Impact Studies

In contrast to resource monitoring programs, the purpose of some studies performed on rocky shores is to detect or quantify the impacts of a stressor on intertidal populations or communities. These studies generally focus on identifying the biological changes that have taken place following site exposure to the impacting agent and the location or geographic extent of the impact. For example, such studies may be designed to determine the effects of improving public access to a rocky shoreline near a populated area, or of a new outfall pipe on the coast, or of a small oil spill on a small section of coastline.

There has been extensive discussion in the literature (Schmitt and Osenberg 1996; Kingsford 1998) of experimental designs to answer these types of questions (see chapter 1). One broad class of analytical methods is the before–after–control–impact (BACI) design or non-BACI post-impact design (e.g., Underwood 1994; Wiens and Parker 1995; Osenberg et al. 1996; Stewart-Oaten et al. 1992; Schmitt and Osenberg 1996). These designs rely on finding suitable impacted and nonimpacted (control) sites and, optimally, on being able to gather data at all sites both before and after the impact. Statistically powerful data are those that show temporal coherence among sites; that is, in the absence of an impact, all sites show similar trends. This outcome in turn relies on the sites being initially similar both in their biota and in their temporal dynamics. In practice, such designs often are asymmetrical and include one impacted site and several randomly selected control sites, with the latter serving to assess natural patterns of temporal and spatial variability (Underwood 1994). Thus selecting appropriate control sites to compare with impacted sites is a critical decision, as is surveying enough sites to be able to factor out natural site-to-site variation.

An alternative to control–impact designs is to establish sites along a gradient of disturbance or potential impact. Such designs can be particularly powerful where a disturbance attenuates with distance from a point source, such as around oil drilling operations, around sewage outfalls, or along a gradient of spilled oil. Using regression analyses, Ellis and Schneider (1997) compared the power to detect change of a control–impact design with a gradient impact design around a well-studied oil field in the North Sea. With equal numbers of observations for each analysis, they found that the gradient impact design was more powerful at detecting changes in abundances of benthic organisms than the control–impact design. Additional advantages of gradient designs are that (1) there is no need for potentially arbitrary selection of control sites (how far away from a disturbance should a control site be located?); (2) gradient analyses do not rely on data that must meet the requirements and assumptions for use of analysis of variance (ANOVA) or similar statistical tests; and (3) they provide information on the spatial scale of the impact, which is useful both for ecological analyses and for management and policy purposes. Ellis and Schneider (1997) caution, however, that similarly to when using control–impact designs, investigators must be certain that all the sites are "comparable with respect to physical processes," so that a gradient in a factor such as depth or sediment grain size is not mistaken for an impact of the pollutant. Thus, physically matching sites for gradient analyses is as critical as it is for other impact-detection protocols.

Ideally, for either type of impact-detection study, we might want to have biotic surveys of many sites and then choose the ones that are most similar overall in "before" community composition. However, such surveys cost time and money and useful prior data are almost never available. Thus, site selection may have to depend on characteristics that are easier to quantify than biotic similarity (and biotic similarity alone may be misleading because of the possibility of different dynamics among sites). Geographic proximity is likely to be a very important parameter in choosing sites for impact detection, because (all else being equal) closer sites are more likely to have more similar biological characteristics and dynamics than broadly separated sites. Other parameters to be "matched" in site selection procedures have been described above.

NUMBERS OF SAMPLING SITES

Inevitably, choices have to be made between extensive sampling (many sites spread over a broad geographic area) and intensive sampling (few sites, but sampled with greater replication in space or time). How many sites should be monitored? Urquhart et al. (1998, 255) note that "where among-site variation across the region of interest is even moderate, a single site produces little data of relevance for inferences across the set of sites." They add that multiple sites help little if there was bias in selecting them. Lewis (1980) notes that sampling only a few sites "runs the high risk of the data being atypical and, therefore, of little value for reaching broader conclusions." In addition, sampling over several sites encompassing a greater spatial scale is more likely to capture community stability (i.e., natural increases and decreases in populations across several sites may balance each other out [Connell 1986]), and managers are less likely to initiate actions based on changes seen at one or a few sites. Spreading sites over a large region also makes it more likely that one will be near a site where, for example, oil strands following a spill, because spilled oil is notoriously patchy.

A useful approach to determining the number of sites that are required for an effective monitoring program may be to combine many of the advantages of a "few"-site with those of a "many"-site protocol. In this type of program, a broad network of sites is established and investigated at a moderate sampling intensity together with a few sites that are intensively sampled, with the smaller set of sites being chosen in such a way as to allow inference of process-level information to the broader region. If trends in measured biological parameters at these sites are coherent both through time and across sites, then the identified biological patterns can serve as particularly powerful templates against which changes due to an environmental stressor or natural shifts in physical conditions can be detected.

In contrast, sampling for impact determination is almost always intensive and focused on a few impacted sites and nearby control sites or on a series of sites along an impact gradient. For either extensive or intensive sampling designs, obtaining statistically and ecologically valid data on impacts will depend on (1) finding sites that are geophysically matched as closely as possible and (2) replicating sufficiently for each class of sites. Underwood and Chapman (1998a, 1998b) emphasize the very high variation in biological parameters among sites ("shores") that they considered to be physically similar (e.g., all wave-exposed), highlighting the need for replication at the site level to obtain sufficient power to detect patterns (see also Dethier and Schoch 2005). While their sites may not have been geophysically matched (or stratified) as closely as recommended here, the point is well taken that data on natural among-site variation must be available before "among-treatment" variation can be ascribed to anthropogenic causes. Thus, site-level replication is a crucial component of any robust study design.

SUMMARY

As stated by Ellis and Schneider (1997), numerous investigators have noted that "more than any other factor, our inability to explain natural variability places a limit on our ability to detect anthropogenic change." Because differences in physical variables among sites increase this natural variability, careful site selection is a critical component for study programs designed to gain understanding of regional trends or to assess impacts on rocky intertidal populations and communities. If sites are not randomly selected, valid (statistical) inferences about other sites of interest cannot be made. If attention is not paid to matching site characteristics, then we run substantial risk of confounding impact detection with natural among-site variation in biotic features caused by subtle geophysical differences. Quantifying impact or discussing trends in biota through space and time requires that site selection (or at least primary site selection) is statistically random (e.g., potential sites are drawn from detailed maps of shorelines) and that geophysical characters important to community structure are matched or at least recorded.

Site-selection protocols (see fig. 2.1) should thus involve (1) identifying a set of sites that are similar in geophysical features and types of habitats to each other or to a disturbed or impacted location of interest and then (2) randomly choosing a subset for study from this site pool (Glasby and Underwood 1998). In addition, sampling should be concentrated on habitats of high value (to humans or to the ecosystem) or of high sensitivity to suspected stressors and, if it is useful to consider trends over a large region, on habitats that are broadly present. Microhabitats such as tidepools and crevices that are individually unique are less likely to be useful for sampling programs that seek to understand a larger area of shoreline. In designing monitoring programs, a balance must be struck between intensive sampling of a few sites and extensive sampling of many sites so that a better picture of biological trends and dynamics can be attained over a broader region. This approach argues for establishing a broad network of sites that are sampled at a lower intensity (perhaps only in one zone or for only a few key species; see chapter 3), combined with a few sentinel sites that receive intensive sampling. Besides providing process-level information that may apply throughout the broader region, such sentinel sites can also provide valuable data about the components of variability (e.g., within sites, among sites, and among years), which in turn can help interpret patterns seen over larger spatial and temporal scales.

LITERATURE CITED

Addessi, L. 1995. Human disturbance and long-term changes on a rocky intertidal community. *Ecol. Appl.* 4:786–97.

Airoldi, L., and F. Cinelli. 1997. Effects of sedimentation on subtidal macroalgal assemblages: an experimental study from a Mediterranean rocky shore. *J. Exp. Mar. Biol. Ecol.* 215:269–88.

Airoldi, L., and M. Virgilio. 1998. Responses of turf-forming algae to spatial variations in the deposition of sediments. *Mar. Ecol. Progr. Ser.* 165:271–82.

Archambault, P., and E. Bourget. 1996. Scales of coastal heterogeneity and benthic intertidal species richness, diversity and abundance. *Mar. Ecol. Progr. Ser.* 136:111–21.

Ardisson, P. L., and E. Bourget. 1997. A study of the relationship between freshwater runoff and benthos abundance: a scale-oriented approach. *Est. Coastal Shelf Sci.* 45:535–45.

Beck, M. W. 1997. Inference and generality in ecology: current problems and an experimental solution. *Oikos* 78:265–73.

Bell, E.C., and M.W. Denny. 1994. Quantifying "wave exposure": a simple device for recording maximum velocity and results of its use at several field sites. *J. Exp. Mar. Biol. Ecol.* 181:9–29.
Bowman, R.S. 1978. Dounreay oil spill: major implications of a minor incident. *Mar. Pollut. Bull.* 9:269–73.
Brosnan, D.M., and L.L. Crumrine. 1994. Effects of human trampling on marine rocky shore communities. *J. Exp. Mar. Biol. Ecol.* 177:79–97.
Castilla, J.C., D.K. Steinmiller, and C.J. Pacheco. 1998. Quantifying wave exposure daily and hourly on the intertidal rocky shore of central Chile. *Rev. Chil. Hist. Nat.* 71:19–25.
Christie, H. 1985. Ecological monitoring strategy with special reference to a rocky subtidal programme. *Mar. Pollut. Bull.* 16:232–35.
Coastal Engineering Research Center (CERC). 1984. *Shore Protection Manual, Vol. 3.* U.S. Army Corps of Engineers, Vicksburg, MS.
Connell, J.H. 1986. Variation and persistence of rocky shore populations. In *The ecology of rocky coasts*, ed. P.G. Moore and R. Seed, 57–69. New York: Columbia Univ. Press.
Craik, G.J.S. 1980. Simple method for measuring the relative scouring of intertidal areas. *Mar. Biol.* 59:257–60.
Dayton, P.K. 1971. Competition, disturbance, and community organization: the provision and subsequent utilization of space in a rocky intertidal community. *Ecol. Monogr.* 41:351–88.
Denny, M.W. 1985. Wave forces on intertidal organisms: a case study. *Limnol. Oceanogr.* 30:1171–87.
Dethier, M.N., and G.C. Schoch. 2004. The consequences of scale: assessing the distribution of benthic populations in a complex estuarine fjord. *Estuarine Coast. Shelf Sci.* 62:253–270.
Doty, M.S. 1971. Measurement of water movement in reference to benthic algal growth. *Bot. Mar.* 14:32–35.
Ellis, J.I., and D.C. Schneider. 1997. Evaluation of a gradient sampling design for environmental impact assessment. *Environ. Monit. Assess.* 48:157–72.
Glasby, T.M., and A.J. Underwood. 1998. Determining positions for control locations in environmental studies of estuarine marinas. *Mar. Ecol. Progr. Ser.* 171:1–14.
Griffiths, C.L., and G.M. Branch. 1997. The exploitation of coastal invertebrates and seaweeds in South Africa: historical trends, ecological impacts and implications for management. *Trans. Roy. Soc. S. Afr.* 52:121–48.
Harley, C.D.G., and B.S.T. Helmuth. 2003. Local and regional scale effects of wave exposure, thermal stress, and absolute vs. effective shore level on patterns of intertidal zonation. *Limnol. Oceanogr.* 48:1498–1508.
Harper, J.R., D.E. Howes, and P.D. Reimer. 1991. Shore-zone mapping system for use in sensitivity mapping and shoreline countermeasures. In *Proceedings of the 14th Arctic and Marine Oilspill Program (AMOP), Environment Canada*, 509–23.
Howes, D., J. Harper, and E. Owens. 1994. *British Columbia physical shore-zone mapping system.* British Columbia Ministry of the Environment, Victoria, Canada.
Jassby, A.D. 1998. Interannual variability at three inland water sites: implications for sentinel ecosystems. *Ecol. Appl.* 8:277–87.

Kiirikki, M. 1996. Mechanisms affecting macroalgal zonation in the northern Baltic Sea. *Eur. J. Phycol.* 31:225–32.

Kingsford, M.J. 1998. Analytical aspects of sampling design. In *Studying temperate marine environments. A handbook for ecologists,* ed. M. Kingsford and C. Battershill, 49–83. Christchurch, New Zealand: Canterbury University Press.

Lewis, J. R. 1980. Objectives in littoral ecology—a personal viewpoint. In *The shore environment. Vol. I. Methods,* ed. J. H. Price, D. E. G. Irvine, and W. F. Farnham, 1–18. London: Academic Press.

———.1982. The composition and functioning of benthic ecosystems in relation to the assessment of long-term effects of oil pollution. *Philos. Trans. R. Soc. Lond. Ser. B Biol. Sci.* 297:257–67.

Littler, M. M. 1980. Overview of the rocky intertidal systems of southern California. In *the California Islands: proceedings of a multidisciplinary symposium,* ed. D. M. Power 265–306. Santa Barbara, CA: Santa Barbara. Museum of Natural History.

Littler, M. M., and D. S. Littler. 1987. Rocky intertidal aerial survey methods utilizing helicopters. *Rev. Photo-Interpret.* 1987–1 (6):31–35.

Littler, M. M., D. R. Martz, and D.S. Littler. 1983. Effects of recurrent sand deposition on rocky intertidal organisms: importance of substrate heterogeneity in a fluctuating environment. *Mar. Ecol. Progr. Ser.* 11:129–39.

Littler, M. M., D. S. Littler, S. N. Murray, and R. R. Seapy. 1991. Southern California rocky intertidal ecosystems. In *Ecosystems of the world, Vol. 24. Intertidal and littoral ecosystems,* ed. A. C. Mathieson and P. H. Nienhuis, 273–96. Amsterdam: Elsevier.

Lohse, D. P. 1993. The effects of substratum type on the population dynamics of three common intertidal animals. *J. Exp. Mar. Biol. Ecol.* 173:133–54.

Luckhurst, B. E., and K. Luckhurst. 1978. Analysis of the influence of substratum variables on coral reef communities. *Mar. Biol.* 49:317–23.

McCook, L. J., and A. R. O. Chapman 1997. Patterns and variations in natural succession following massive ice-scour of a rocky intertidal seashore. *J. Exp. Mar. Biol. Ecol.* 214:121–47.

McDonald, L. L., W.P. Erickson, and M. D. Strickland. 1995. Survey design, statistical analysis, and basis for statistical inferences in coastal habitat injury assessment: *Exxon Valdez* oil spill. *Exxon Valdez oil spill: fate and effects in Alaskan waters.* ASTM STP 1219, ed. P. G. Wells, J. N. Butler, and J. S. Hughes, 296–311. Philadelphia, PA: American Society for Testing and Materials.

Menge, B. A., B. A. Daley, P. A. Wheeler, and P. T. Strub. 1997. Rocky intertidal oceanography: an association between community structure and nearshore phytoplankton concentration. *Limnol. Oceanogr.* 42:57–66.

Murray, S.N., and R.N. Bray. 1993. Benthic macrophytes. In *Ecology of the Southern California Bight,* ed. M. D. Dailey, D. J. Reish, and J. W. Anderson, 304–68. Berkeley: University of California Press.

Murray S. N., T. G. Denis, J. S. Kido, and J. R. Smith. 1999. Frequency and potential impacts of human collecting in rocky intertidal habitats in southern California marine reserves. *CalCOFI Rep.* 40:100–106.

National Research Council (NRC). 1995. *Review of EPA's Environmental Monitoring and Assessment Program: overall evaluation.* Washington, DC: National Academy Press.

Olsen, T. B. P. Hayden, A. M. Ellison, G. W. Oehlert, and S. R. Esterby. 1997. Ecological resource and monitoring: change and trend detection workshop report. *Bull. Ecol. Soc. Am.* 78:11–13.

Osenberg, C. W., R. J. Schmitt, S. J. Holbrook, K. Abu-Saba, and A. R. Flegal. 1996. Detection of environmental impacts: natural variability, effect size, and power analysis. In Schmitt and Osenberg, 83–107.

Page, D. S., E. S. Gilfillan, P. D. Boehm, and E. J. Harner. 1995. Shoreline ecology program for Prince William Sound, Alaska, following the *Exxon Valdez* oil spill: Part 1—Study design and methods. *Exxon Valdez oil spill: fate and effects in Alaskan waters*. ASTM STP 1219, ed. P. G. Wells, J. N. Butler, and J. S. Hughes, 263–95. Philadelphia, PA: American Society for Testing and Materials.

Pugh, P. J. A., and J. Davenport. 1997. Colonization vs. disturbance: the effects of sustained ice-scouring on intertidal communities. *J. Exp. Mar. Biol. Ecol.* 210:1–21.

Raffaelli, D., and S. Hawkins. 1996. *Intertidal ecology*. London: Chapman and Hall.

Raimondi, P. T. 1988. Rock type affects settlement, recruitment, and zonation of the barnacle *Chthamalus anisopoma* Pilsbury. *J. Exp. Mar. Biol. Ecol.* 123:253–67.

Research Planning, Inc. (RPI). 1990. *Environmental sensitivity index maps*. Seattle, WA: National Oceanic and Atmospheric Administration, Office of Ocean Resources Conservation and Assessment Columbia, SC: RPI.

Ricketts, E. F., J. Calvin, J. W. Hedgpeth, and D. W. Phillips. 1985. *Between Pacific tides*. 5th ed. Stanford, CA: Stanford University Press.

Schmitt, R. J., and C. W. Osenberg, eds. 1996. *Detecting ecological impacts: concepts and applications in coastal habitats*. New York: Academic Press.

Schoch, G. C. 1996. *The classification of nearshore habitats: a spatial distribution model*. M.S. thesis. Corvallis: Oregon State University.

Schoch, G. C., and M. N. Dethier. 1996. Scaling up: the statistical linkage between organismal abundance and geomorphology on rocky intertidal shorelines. *J. Exp. Mar. Biol. Ecol.* 201:37–72.

———. 1997. Analysis of shoreline classification and bio-physical data for Carr Inlet (FY97–078, Task 1 and Task 2). Report to the Washington State Dept. of Natural Resources, Dec.

Stephenson, W. 1961. Experimental studies on the ecology of intertidal environments at Heron Island: II. Effects of substratum. *Aust. J. Mar. Freshwater Res.* 12:164–76.

Stewart-Oaten, A., J. R. Bence, and C. W. Osenberg, 1992. Assessing effects of unreplicated perturbations: no simple solutions. *Ecology* 73:1396–1404.

Underwood, A. J. 1994. On beyond BACI: sampling designs that might reliably detect environmental disturbances. *Ecol. Appl.* 4:3–15.

Underwood, A. J., and M. G. Chapman. 1989. Experimental analyses of the influences of topography of the substratum on movements and density of an intertidal snail, *Littorina unifasciata*. *J. Exp. Mar. Biol. Ecol.* 134:175–96.

———. 1998a. Spatial analysis of intertidal assemblages on sheltered rocky shores. *Aust. J. Ecol* 23:138–57.

———. 1998b. Variation in algal assemblages on wave-exposed rocky shores in New South Wales. *Aust. J. Mar. Freshwater Res.* 49:241–54.

Urquhart, N. S., S. G. Paulsen, and D. P. Larsen. 1998. Monitoring for policy-relevant regional trends over time. *Ecol. Appl.* 8:246–57.
Wethey, D. S. 1985. Catastrophe, extinction, and species diversity: a rocky intertidal example. *Ecology* 66:445–56.
Wiens, J. A., and K. R. Parker. 1995. Analyzing the effects of accidental environmental impacts: approaches and assumptions. *Ecol. Appl.* 5:1069–83.
Witman, J. D., and K. R. Grange. 1998. Links between rain, salinity, and predation in a rocky subtidal community. *Ecology* 79:2429–47.

Researchers (R.T. Paine and R.S. Steneck) studying coralline algae beneath intertidal kelps at Tatoosh Island, Washington State.

CHAPTER 3

Biological Units

The biological units targeted by a sampling program can vary from individual-based parameters such as the size of a particular limpet species' gonads, to population-level parameters such as counts of all macroscopic organisms, to higher taxonomic units such as the numbers of phyla (fig. 3.1). The biological units to be selected will vary (as always) with the goals of the sampling program and with available knowledge about the ecology of the populations and communities being studied. The investigator will want to choose the most informative biological units—that is, those that best address the goals of the study and that have the potential to provide statistically powerful answers to the specific research questions. Ideally, the chosen biological units also will have known causal links with any stressors being studied. Unfortunately, rarely are clear causal relationships between stressors and responses of rocky shore organisms known, even when only a single identifiable stressor is under consideration.

SPECIES-LEVEL SAMPLING

Sampling the abundances of one or more species populations at target sites is a common component of most rocky shore sampling programs. There are several advantages to species-level sampling. (1) The species, or local population, is the basic unit of the majority of ecological research; ecologists are comfortable thinking in terms of species populations, and there are established sampling designs and procedures for determining the abundances of species varying in size and distribution. (2) Researchers usually explore the ecological processes affecting communities at the species level. Thus, background information on trophic position, life history, recruitment patterns, and other ecologically important topics is

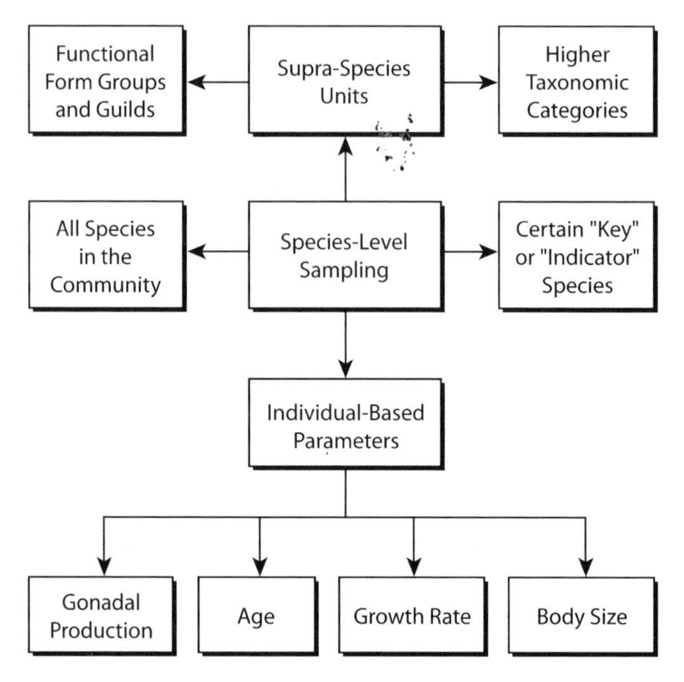

Figure 3.1. Choices of biological units in rocky intertidal sampling programs. Obtaining data on species abundances is often the focus of intertidal sampling programs. Data can be collected for all species in the habitat or for selected "key" or "indicator" species. Species data can be assigned to higher taxonomic categories or to functional groups but collecting data only at supraspecies categories limits options because data for individual species cannot be reconstituted. Down-a-level sampling programs focus on individual-based parameters that may be more sensitive to environmental change than the abundances of species or functional groups.

often available for the most common coastal species. (3) The abundances of populations, whether measured as density, percentage cover, or biomass (see chapters 6 and 7), presumably integrate and reflect local environmental conditions operating at the study site. Individual-based parameters such as gonadal production may be more sensitive to unusually good or bad environmental conditions (e.g., exposure to pollutants, nutrient availability, or changes in predation pressure), but changes in environmental factors should ultimately be manifest directly or indirectly in the size of a population. Moreover, abundance responses in populations can usually be measured over short time scales when abrupt changes in abiotic or biotic conditions significantly increase mortality rates. However, resistant and long-lived species might not show immediate abundance shifts but, instead, may show effects over longer time scales due to reduced larval production or recruitment. (4) Even when sampling is not targeted toward determining the effect of a particular stressor, if abundance data are collected on a diversity of species types (e.g., long and short-lived species, producers and consumers, or species with different dispersal shadows), then we should be able to detect changes in one of these populations that correspond with the onset of a stressor. (5) Alternatively, by focusing on a limited number of ecologically important species, investigators need only recognize the species of interest, and data collection will not require identification of all of the taxa found in a community or assemblage.

Several disadvantages of species-level sampling programs exist. These include designing sampling strategies that account for natural variability at the population level in both space and time and the inherent difficulties in identifying *a priori* the appropriate species to sample. Numerous studies have demonstrated high levels of variation in the abundances of rocky intertidal plant and animal populations due to spatial patchiness and temporal dynamics. This variation can be driven by natural abiotic parameters, such as changing water temperatures, exposure to large storm waves, or fluxes in sand movement, and biotic processes including spatially variable patterns of predation and recruitment. Such high noise-to-signal ratios can make it particularly difficult to statistically detect changes at the population level given the limits of most sampling programs. Even carefully designed before–after–control–impact (BACI) sampling programs (see chapter 1) can become uninformative when species abundances vary greatly among sites or over time. Likewise, if the compared sites track different temporal courses, there will be a significant site × time interaction resulting from factors unrelated to the stressor of interest. For example, Osenberg et al. (1994) found a large effect size (i.e., a large impact) on population-level abundance data in response to a pollutant; however, this effect was very difficult to detect statistically because of high within- and among-site natural variability (i.e., high noise-to-signal ratio) in the abundance data.

The problem of high natural noise levels has led to concerns that a truly informative, rocky intertidal monitoring program relying on species abundance data cannot be developed. However, the degree of among-site variation in biological parameters can be reduced through careful selection of the sites and populations to be studied (chapter 2) and the employment of appropriate and well-crafted sampling designs (chapters 4 and 5;). Lewis (1982, 260) notes: "While our objective is the detection of either a unidirectional trend or a persistent departure from normal, the cycles and fluctuations that constitute the normal make the task difficult. But at least they can be recognized for what they are—by an experienced field naturalist—and many can be discounted...." For all of these reasons, selection of the biological units to be sampled is a critical component of any rocky intertidal research program. This decision affects the ability to detect differences between treatments and controls or over spatial and temporal scales and, ultimately, to gain understanding of causal links between environmental and population parameters.

Unfortunately, opinions about "what species to sample" are almost as diverse as rocky intertidal researchers. Generally, the tendency is to identify key species or, as stated by Clarke (1997), "those whose presence or absence in certain environments has major repercussions in the community." If a study is designed to detect present or future effects of a particular stressor (e.g., oil spills, sewage effluents) or management action (e.g., establishing reserve protection), then so-called "indicator species" might be appropriate to sample. These should be taxa known to be particularly resistant (e.g., the polychaete *Capitella capitata,* which is abundant around sewage discharge sites) or sensitive (e.g., the bivalve *Brachidontes rodriguezi,* whose numbers decline in areas highly affected by sewage effluents [López Gappa et al. 1990, 1993]) to the stressor or action. However, few species that are good "biomonitors" or "indicators" of particular stressors have been unequivocally identified for rocky intertidal systems. Requirements for a good species for biomonitoring are reviewed by Underwood and Peterson (1988), Keough and Quinn (1991), and Loeb and Spacie (1994); these include (1) wide distributions to allow for comparisons among geographic areas, (2) a well-known autecology to reduce the chances of reaching false conclusions if changes are detected by the sampling program, (3) sensitivity to the stressor of concern but not to variations in other environmental conditions that likely occur over the study area, (4) the ability to be sampled with precision and accuracy, and (5) the capacity to serve as a representative of a wider range of species or ecological processes.

In addition, political and public concerns may enter into the choice of target species and influence the design of monitoring or impact studies. For example, there may be pressure to monitor certain "charismatic" species or species of high conservation interest, such as sea or shore birds or marine mammals such as sea otters, or species of economical value. Identifying a species that fulfills these requirements, while also meeting the criteria for a good biomonitor or indicator, may not be possible, especially when species selection is of secondary importance to site selection in a region. In these cases, if sites are selected randomly, not all chosen sites may contain suitable populations of a targeted indicator species. Underwood and Peterson (1988) also caution against choosing a "millstone" species (i.e., a species selected because of its historical use), instead of a keystone species (*sensu* Paine 1966) known to play an important role in community organization. Selection of politically expedient, charismatic or "millstone" species to serve as the basis for a research or monitoring program may jeopardize the scientific value of the program at the outset unless the program goal is simply to monitor the abundance of that special species.

Often researchers are asked to design sampling programs for monitoring spatial and temporal changes in biota in anticipation of a future unidentified perturbation. However, the *a priori* selection of appropriate biomonitors for such programs may not be possible because the conditions or stressors of concern are likely unknown and multiple. Therefore, a species selected as a good indicator for detecting sewage pollution may not be good for assessing the effects of human foot traffic or overexploitation. As a consequence, when a single comprehensive study is desired, investigators may hedge their bets by selecting one or more species thought to play key ecological roles in the ecosystem or by sampling all macroscopic organisms in the community.

Lewis (1982) has argued vigorously for focusing sampling on populations of the dominant structure-forming species (e.g., mussels, surfgrass) and key grazers and predators in the community because most other rocky shore organisms will depend on these "key" species. Thus, other unstudied species populations should show patterns of change that correlate with changes in the key species. Many dominant or ecologically important species in rocky intertidal communities are large, long-lived, and easy to identify, offering the additional advantage of being simpler to sample with accuracy and precision. Houghton et al. (1997), for example, found that sampling four dominant taxa (*Fucus*, Lottiid limpets, *Littorina*, and *Nucella*) provided most of the information needed to detect functional changes in rocky intertidal communities on sheltered shores in Prince William Sound, Alaska, following the *Exxon Valdez* oil spill. Similarly, Ambrose et al. (1994; see also Engle et al. 1994) selected 10 ecologically important and well-understood species or species groups (e.g., mytilid mussels, acorn and gooseneck barnacles, *Anthopleura* spp., *Silvetia* [= *Pelvetia*]) as the basis for a monitoring program designed to enable detection of the effects of a major disturbance, such as a large oil spill, on California shores. Long-lived or ecologically dominant species, however, might not respond to changing environmental conditions or stress as rapidly as short-lived annuals or less common species. In addition, some ecologically important species, such as sea stars, may be too mobile and patchy in their distributions for designing efficient sampling programs that can statistically track changes in their densities over time.

The uncertainties associated with knowing which species to sample have led other researchers to focus their sampling efforts on all or most species in an intertidal community. The arguments for this community-level approach are that

(1) a single "canary-in-the-mine" indicator species for rocky shore ecosystems has yet to be identified;
(2) with few exceptions, how changes in the abundances of key species affect other species is unknown in rocky intertidal communities;
(3) rare species may be more sensitive to changing environmental conditions or the presence of pollutants than more common, well-studied ones;
(4) sampling more than one species presents the opportunity to examine coherence in trends among multiple components of the community, which in turn strengthens conclusions about the effects of changing conditions or impacts (e.g., Schroeter et al. 1993); and
(5) gathering data on only a few species compromises the ability to examine changes in community parameters and to perform useful assemblage-level multivariate analyses.

Rare species seem to be particularly numerous on rocky shores and present significant problems because by definition they will show up in only a few sampling units and will be difficult to sample with accuracy and precision. In addition, their scarcity presents difficulties during data analysis because they usually exhibit high variation in abundance among replicate samples and the numerical data describing their distributions and abundances through space and time are zero-rich, violating assumptions of most univariate and multivariate statistical tests. Ignoring rare species, however, also presents the risk of excluding taxa that might be highly responsive to changing environmental conditions, the presence of stressors, or experimental treatments. The debates on the value of rare species in sampling programs designed to detect ecologically meaningful changes in aquatic communities are ongoing (Cao et al. 1998; Marchant 1999; Cao and Williams 1999). Clearly, more research is needed to improve our understanding of the ecological roles of many of these rare species in intertidal and other aquatic ecosystems, including their resistance and sensitivity to disturbances and changing environmental conditions. One analytical solution (L. Tear, pers. comm.) is to treat rare species differently from species for which it is possible to collect quantitative data. Criteria for determining rare versus common species can be set and the species composition of these two groups can be compared separately between times and locations. It may be more useful to note which species are "rare" or "common" and how those lists change at a site over time or between sites than to try to find a metric and a method of analysis appropriate for a combined dataset that includes both species types.

Although comprehensive in scope, sampling programs that track all species in a rocky intertidal community have major disadvantages. Principal among these are that (1) the observers performing the fieldwork must have the taxonomic expertise to discriminate all encountered taxa, (2) much more time and cost will be required to carry out field sampling and data reduction and analysis tasks compared with studies that focus on only a few select species, and (3) no single sampling design or method can adequately assess all species in an intertidal community. Different approaches are required to optimize sampling of mobile and sessile species, abundant and scarce species, and large and small species, resulting in the need for complex, mixed sampling designs when carrying out comprehensive sampling programs. If only a few key populations are targeted, then it might be feasible to employ separate sampling strategies for each population of interest to best capture information on their abundances (e.g., for common vs. rare species).

Under circumstances where prior data are available, either over several sites or over several years, an informative approach is to focus data gathering on species that show the lowest spatial or temporal variation (e.g., lower error-to-mean ratios). This strategy can reduce among-sample variation or noise and increase statistical power when attempts are made to detect differences or changes between sets of samples. Ideally, the populations chosen for focused sampling will include a diversity of species types (e.g., based on their population abundances, trophic positions in the community, and life histories). For example, on the Olympic coast of Washington, candidate species for such a focused sampling program include barnacles that occupy the very high intertidal zone, certain limpets, abundant seaweeds in wave-protected habitats such as *Fucus gardneri* and *Endocladia muricata*, the anemone *Anthopleura elegantissima*, and crustose coralline algae (Dethier, unpubl. data). However, these species may not show this same low variability among sampling units in other geographic regions. Data obtained from such a sampling program can be augmented by less frequent, intense sampling that includes all species populations in the community, perhaps even as simple checklists. These data could allow community-level comparisons of parameters such as richness and diversity and facilitate the use of multivariate techniques that may help establish relationships between changes in species populations and variations in treatments or environmental conditions.

Additional data that may be informative to gather at the species level include measurements of the distributions of species populations, either over short horizontal spatial scales or across the entire intertidal zone. Because almost all intertidal plants and animals are restricted in their distributions to particular tidal zones, changes in their vertical distributions may provide ecologically important information. For example, a gradual rise in the elevation of a barnacle band on the shore might suggest a rise in sea level or a reduction in exposure to desiccating conditions. Abrupt loss of the top of an intertidal algal band corresponding with an oil spill might provide a useful measure of impact, even if the individuals found in the central part of the species distribution remain healthy. Loss of high–shore mussels on north- but not south-facing shores on the coast of Washington, for example, helped to differentiate losses from a near-simultaneous oil spill and a major freeze event (Dethier 1991). Often such distributional information can be obtained simply and accurately, and at the needed scale, by taking a series of photographs during each site visit using a consistent reference point. Shifts in distributions over short horizontal scales within the same vertical tidal range also may be important to document. This will be particularly true for long-term monitoring programs that employ fixed plots because species abundances can decrease or increase inside plots but show the opposite effects in adjacent, unsampled areas. Such situations can lead to misleading conclusions if comparisons among sites or over time are based exclusively on data taken from the fixed plots and leave undetected real changes in patch size, shape, and distribution. Some of the advantages and disadvantages of fixed plot sampling are discussed in chapter 4.

DOWN-A-LEVEL: INDIVIDUAL-BASED SAMPLING

Several authors, including Jack Lewis, an experienced marine ecologist who has monitored rocky shore populations for decades (e.g., Lewis 1976; Lewis et al. 1982), argue that measurements of the individual performances of one or more key species at a site are more sensitive indicators of environmental change than abundance data. Indexes of performance include growth rate, gonadal output, certain physiological parameters of individual animals, and the recruitment strength of the population. Thus, Lewis and coworkers (e.g., Heip et al. 1987) have concentrated their monitoring efforts on measuring parameters that provide information about the status of a population, such as size and age structure, reproductive condition, and recruitment strength. Reproductive cycles of various mollusks in the United Kingdom, for example, are highly sensitive to climatic and hydrographic change (studies in Heip et al. 1987). Often, pollutants impact gonadal development and certain biochemical processes, and when these connections are known, measures of such individual species' performances can be used as powerful biomonitors. Zeh et al. (1981) also have stressed the value of measuring size structures of populations through time, because these data provide information on a population's development and serve as a sensitive means for detecting change in population status. Osenberg et al. (1994) argue for following body size and other individual-based parameters because they found that these were generally more consistent through time at control sites and more variable at impacted sites. They concluded that the statistical power to detect change in these parameters exceeded the power available using population-based parameters such as density, where data were more variable among sites and times.

A possible disadvantage of relying on individual-based parameters is that the chosen measures may vary greatly among sites, to the extent that among-site differences may not be distinguishable from differences resulting from large-scale disturbances. Gathering performance data also is time-consuming and may require destructive sampling (e.g., to measure gonadal sizes) and careful timing (e.g., to obtain gravid specimens when reproduction is highly seasonal). For instance, growth rates in at least some intertidal invertebrates are so responsive to local environmental conditions (e.g., population density, competition, availability of food resources) that they cannot be used as an index of broader regional patterns. Likewise, other performance parameters (such as recruitment, maximum body size, or population age structure) may vary greatly over local spatial scales or due to different site histories. If parameters are consistent enough within a site to establish a baseline (as in the case described by Osenberg et al. 1994), however, then changes through time may be important indicators of changing environmental conditions or of site disturbance. Before such individual-based performance data become the focus of any study, more information is needed to determine which species are good and sensitive integrators of environmental conditions and to establish baseline performance expectations under natural (stressor-free) conditions. Selected approaches to individual-based studies of intertidal seaweeds and macroinvertebrates are discussed in chapter 8.

UP-A-LEVEL: SUPRASPECIES SAMPLING

Because identifying all species in a community requires considerable expertise and time and because some species of ecological importance can be difficult to consistently discriminate and sample, other ways to meaningfully categorize biological units have been sought. Approaches to this categorization have varied depending on the investigator and the habitat being studied. For example, some investigators who work in soft-bottom benthic communities favor grouping sampled biological units into higher taxonomic categories. Alternative approaches used in rocky intertidal and subtidal habitats include categorization of species into guilds or functional groups. Arguments concerning the value of this supraspecies-level sampling vary but generally hinge on trade-offs between the unwanted loss of information and increased potential for obtaining misleading results versus greater sampling efficiency and a savings in study costs.

Higher Taxonomic Categories

Several studies, mostly performed in habitats other than the rocky intertidal zone, have demonstrated that data gathered at higher taxonomic levels (e.g., genus, family, or even division or phylum) can substitute for more time-consuming species-level sampling (Keough and Quinn 1991; Underwood 1996). For example, Kratz et al. (1994) found in alpine lakes that quantifying rotifers at the species level resulted in variable and difficult-to-interpret data. However, when data were grouped by genus they were less variable because of compensatory processes within the aggregation and showed strong effects of lake acidity on rotifer abundances. In sedimentary marine habitats, workers have found that quantifying the abundances of polychaete annelids by family instead of by species effectively describes trends as well as sensitivity to stresses from pollution (e.g., Ferraro and Cole 1992; James et al. 1995; Olsgard et al. 1997). For example, using this approach Warwick (1988a, 1988b) found no substantial loss of sensitivity in detecting patterns at the family versus the species level. He reported that "taxonomic sufficiency is required only to the level that indicates the community response" and that huge savings in time and money can result from gathering data at higher taxonomic levels. A possible functional explanation for this outcome is that variation in natural environmental conditions, such as sediment grain size or water depth, may influence the fauna through species replacement (Warwick 1988a). In contrast, exposure to anthropogenic stressors modifies the community at a higher taxonomic level because species belonging to the same family (or even phylum) are similar in their abilities to respond to the altered conditions. Underwood (1996) notes, however, that the rules for reducing taxonomic effort must be known and that this requires understanding of how supraspecies levels respond to impacts. Unfortunately, mechanistic understanding of why supraspecies analyses work to detect changes in spatial pattern due to disturbance has yet to be developed. Often, data are gathered at the supraspecies level not because of a tested functional similarity in responses among lower-level taxa, but because of limited time or taxonomic expertise (e.g., Morrisey et al. 1992). This illustrates both the advantages and the disadvantages of relying on higher taxonomic categories in sampling programs. Sampling species by lumping them into higher taxonomic groupings is easier, simplifies data analysis, and is, therefore, less costly; however, this approach is often taken without ecological justification, and therefore, the results and conclusions of such studies may lack validity and be misleading.

A cogent example of how sampling higher-level groups can lead to loss of information when studying rocky intertidal populations can be found in recent research on two barnacle species. Suchanek (pers. comm.)

found that two congeners of barnacles had very different sensitivities to *Exxon Valdez* oil in the field. Collecting barnacle abundance data at the genus level would be simpler and more efficient in this situation but would fail to accurately characterize the impact of oil on these species and lead to errors in interpretation. However, it could be argued that changes in the distribution and abundance of a single species are not always ecologically important, whereas changes occurring in all co-occurring congeners or throughout a family are indicative of serious and significant community modifications.

Another disadvantage of gathering data at the supraspecies level is that this approach precludes analyses of biodiversity based on species richness. Lumping information to analyze patterns at higher taxonomic levels can always be done after collecting data at the species level, but discrimination to the species level of resolution can only occur at the time of the initial data gathering. Any study that requires inventories of all species in a community, however, requires that the investigators have extensive taxonomic expertise and results in significant increases in field and laboratory sampling time. Oliver and Beattie (1996) illustrated a potentially useful compromise for sampling arthropod biodiversity and determining species turnover in different forest types. They found that collections sorted by nonspecialists into "morphospecies" (taxa that differ from each other clearly, based on external features) showed diversity patterns and trends similar to those generated by taxonomic specialists working at lower taxonomic levels. They also found that some of these morphospecies could serve as "surrogate taxa" that show strong relationships with other taxa. However, they cautioned against extending this approach to other organisms and habitat types without careful empirical testing. Clearly, research is needed to determine the ecological utility of using lumped species, morphospecies, or surrogate taxa when performing rocky intertidal studies.

Functional Groups

An alternative approach to using systematic criteria for lumping organisms into biological units is to group species with similar functional characteristics such as body form, trophic position ("guild"), or life history pattern (fig. 3.2). The rationale for this approach is that the various species within a functional group (e.g., mobile deposit-feeding worms, fine filamentous algae, microalgae-consuming limpets) may be unpredictably present through space and time, while the abundance of the group as a whole remains relatively stable. For example, different species belonging to a functional group may colonize a rocky intertidal surface at

Functional Groups	Thallus Size (m)	Morphology
1 Microalgae (single cell) Cyanobacteria and diatoms	$10^{-6} - 10^{-5}$	50µm
2 Filamentous Algae (uniseriate) Cladophora and Bangia	$10^{-3} - 10^{-4}$	1mm; 300µm
3 Foliose Algae (single layer) Monostroma or multilayered Ulva, Porphyra	10^{-1}	
3.5 Corticated Foliose Algae Dictyota and Padina	10^{-1}	5cm
4 Corticated Macrophytes (terete) Chondrus and Chondracanthus	10^{-1}	5cm
5 Leathery Macrophytes Laminaria and other Kelps and Fucus	$10^{-1} - 10^{-1}$	10cm; 50cm
	Anatomical Morphological Continuum Increasing Complexity & Thallus Size (↓ groups 1–5)	
6 Articulated Calcareous Algae Corallina and Halimeda	10^{-1}	1cm
7 Crustose Algae Lithothamnion, Peyssonnelia and "Ralfsia"	10^{-1}	3cm
	Increasing Complexity & Thallus Size (↑ groups 7–6)	

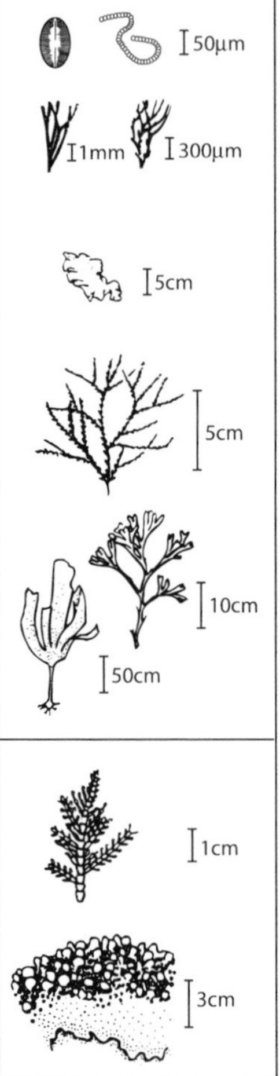

Figure 3.2. Diagrammatic representation of a scheme for classifying seaweeds into seven functional groups. The functional groups (specific morphological form in parentheses), range of thallus sizes, and morphologies of representative genera are presented for each group. (Modified from Steneck and Dethier 1994.)

different places and times, but generally their responses to major ecological processes will be similar. Functional groups can be those that respond to the environment in a similar way (response groups) or those that have similar effects on the rest of their communities (effect groups).

Species that use similar resources in a similar manner should be indicative of the processes that control those resources (Underwood and

Petraitis 1993). These relationships have been demonstrated repeatedly for soft-sediment infauna (e.g., W. Wilson [1991] and Posey et al. 1995; but see Weinberg [1984] for an exception), epibionts growing on mangrove roots (Farnsworth and Ellison 1996), and algae occurring on rocky substrata (reviewed in Steneck and Dethier 1994; also Hixon and Brostoff 1996). In addition, macroalgal primary productivities and other ecological attributes have been found to be associated with functional-form groups (Littler and Littler 1980; Littler and Arnold 1982). Upright, canopy-forming algae, for example, are more susceptible to damage from human foot traffic than low-lying turfs or crusts and thus, as a functional group, constitute a strong indicator of trampling disturbance (Povey and Keough 1991; Brosnan and Crumrine 1994; but see Fletcher and Frid 1996). In addition, ephemeral green algae often replace larger, brown seaweeds and become more abundant in polluted habitats (Keough and Quinn 1991). Functional-group approaches also have been used successfully (Littler and Littler 1984; Murray and Littler 1984; Steneck and Dethier 1994) to analyze macrophyte patterns on rocky shores. If the study goals are to determine the stability of an ecological community, or the effects of large-scale ecological forces, data analysis at the level of functional groups may be useful and informative (Hay 1994). For example, these kinds of data would clearly and simply describe a change in community composition from larger, longer-lived kelps to smaller, opportunistic algal species. Sampling at this level of taxonomic discrimination is clearly faster, requires less expertise, and is less costly compared with sampling at the species level. However, analyses based on a functional-group approach also can be less able to detect biological effects of changes in environmental conditions (Phillips et al. 1997).

Besides sometimes being less informative, another potential disadvantage of using a functional-group approach in an intertidal sampling program is that, as always, the proper interpretation of patterns or trends depends on knowledge of the natural histories of the sampled organisms. Species within a guild may use one resource (e.g., space) in a similar way but be very different from one another in other ways (Menge et al. 1983; Underwood and Petraitis 1993). For example, a change in dominance from *Mytilus californianus*, a long-lived, competitive species, to its congener, *M. trossulus*, a shorter-lived, smaller, and faster-growing mussel, would not be detected if sampling was restricted to supraspecies or functional-group biological units. It may be common to have such variation in life history characteristics, or even pollution tolerance, among species assigned to a particular functional group since groups usually are based on morphology or trophic position. Better empirical knowledge of the functional equivalence of species is needed before investigators rely too heavily on a functional-group approach when designing intertidal sampling programs. In addition, functional groupings have been based on a wide range of criteria that may differ from study to study so that consistent associations between a particular functional group and its ecological attributes are not always apparent (J. Wilson 1999). The functional-group approach also is clearly coarse-grained; it may detect large-scale changes but may not have the sensitivity to detect early changes in conditions that might be detectable through more fine-grained, species-level sampling programs.

COMMUNITY-LEVEL METRICS

Analyses performed on population-based parameters may not always allow detection of ecologically meaningful changes among sites or over time. Moreover, when changes are detected in population or individual attributes for only one or a few species, the representative nature of these changes may be uncertain. In these cases, or to integrate the trends exhibited in community structure by all or most sampled species, community-level analyses are in order. Clarke (1997) argues that approaches involving the whole assemblage of species in a habitat are more realistic than the study of the fate of only selected species. The search for bioindicator species often has been paralleled by a search for a parameter that describes responses of the entire community in a compact and meaningful way. Karr (1994) notes: "Biological monitoring must seek ecological attributes for which natural variation is low but that are influenced to some significant extent by human actions." The most commonly used community-level metric of this type is diversity, which can be expressed either as species richness (i.e., the number of species or taxa present per unit area) or as an index integrating richness with species abundances in the community. Usually, richness or a diversity index is calculated for each sample and the resultant means are analyzed using univariate statistics to determine temporal or spatial differences among the sampled communities. The popularity of diversity indexes probably stems from their being thought of as indicators of ecosystem well-being (Magurran 1988).

Habitat-specific alternatives to diversity exist, such as the nematode-to-copepod ratio and the annelid pollution index for soft sediments (Keough and Quinn 1991). Karr (1991) used an "index of biological integrity" (IBI) to evaluate the status of fish communities in stream ecosystems affected by human activities. This index integrated parameters addressing attributes of individuals, populations, communities, and ecosystems and included metrics describing species richness and composition, trophic composition, and fish abundances and condition. Others have worked to apply the IBI concept to marine systems, especially estuaries (Weisberg et al. 1997). For example, a variety of parameters, including diversity, abundance of individuals, biomass, and percentage of pollution-indicative taxa were used by Weisberg et al. (1997) to create an integrated descriptive metric for evaluating the status of estuarine habitats. To our knowledge, no such attempts have been made to develop a similar integrated metric for describing the condition of rocky shore communities or ecosystems.

The use of species richness and diversity indexes to describe community status stems from the theoretical relationship between stability and diversity and from simulations suggesting that diversity indexes should have greater statistical power to detect differences among communities than analyses based on single species. Too frequently, univariate analyses performed on population-level data fail to detect ecologically meaningful differences because of inadequate statistical power resulting from high variance among sample units, a problem that can be overcome but at the expense of taking an extreme number of replicates or very large sampling units. A community-level approach based on species richness or diversity, however, may improve the statistical power of univariate tests. Hypothetically, unnatural stresses should cause some species to become locally extinct, thereby lowering species richness or decreasing evenness in the abundances of remaining taxa (if they are not replaced by stress-resistant organisms). Although numerous studies (e.g., Littler and Murray 1975; Weisberg et al. 1997) have shown relationships between species richness or diversity and anthropogenic impacts on aquatic communities, consistent relationships do not always occur. For example, trampling, a type of stressor on rocky shores, does not always affect species richness, even though the abundances of individual taxa may be strongly impacted (Keough and Quinn 1991; but see Fletcher and Frid 1996). Moreover, intermediate levels of disturbance may result in higher species diversity in affected communities compared with those that experience either lower or higher levels of perturbation (e.g., Sousa 1979a, 1979b). As discussed for supraspecies parameters, species diversity should be more stable through time than individual species abundances because of compensatory responses within the community. However, only a few studies (Zeh et al. 1981; Karr 1994; Dethier, unpubl.) have been performed to support this assumption. Another disadvantage of relying on diversity metrics is that diversity is not a "unique" measure, i.e., two sites with very different species richness and relative abundances among species can have the same diversity value. Clearly, before diversity is accepted as a key parameter for describing changes in rocky shore communities due to perturbations, its use must be tested across a gradient of habitat types and impact intensities.

Estimates of species richness depend very strongly on the size and number of sample units, the level of taxonomic expertise of the observers, working conditions, observer fatigue, and external variables such as season or even time of day. Therefore, any study that relies on richness data must be carefully designed to account for these variables. Additionally, there are a myriad of indexes that attempt to quantitatively integrate richness and evenness components of diversity, each with its own set of attributes (Pielou 1975; Magurran 1988), making it difficult to arrive at a universal index of choice. Other potential disadvantages of relying on species richness or diversity measures to demonstrate impacts of disturbances on intertidal communities include

(1) the lack of clear and unambiguous causal links between the index and stressors as discussed previously;
(2) the fact that richness/stressor relationships may be nonlinear so that a change in the measured index may not be detected until changes in the biota have become severe;
(3) the potential for being misled by richness patterns in a selected subset of the resident taxa, because not all taxa may show the same trends or patterns (especially in soft-sediment communities);
(4) uncertainty in determining what constitutes a significant ecological change in terms of diversity measures; and
(5) the lack of acceptance by some resource managers and lawyers of richness as a metric describing community status, because they are accustomed to discussing damages to communities in terms of measured losses in the numbers of particular species.

In ecological terms, it is often the change of the entire community that matters, not simply the change in one or a few species. Changes in species composition and relative abundances of a set of species in an assemblage may provide a strong signal about an environmental factor of interest (e.g., pollution discharges or a change in sea level) because the types and abundances of species that form a community effectively integrate and "record" such events (Philippi et al. 1998). Multivariate analyses provide the tools to look at this bigger picture (Underwood 1996), retain information on all sampled species within a community, and may be able to reduce the noise caused by the responses of individual species. In this sense, different species may act as replicates of each other's responses, or their noise contributions may cancel each other out, improving the ability to detect patterns in overall community response (Faith et al. 1991).

There is a bewildering variety of multivariate techniques of potential use in monitoring or impact detection programs or for describing and analyzing spatial and temporal changes in intertidal communities (e.g., see Volume 172 of *Journal of Experimental Marine Biology and Ecology, Changes in Marine Communities,* dedicated to this topic). Legendre and Legendre (1983) review these types of analyses and their appropriate uses. Although other multivariate approaches can provide valuable analysis of community-level data in marine systems (e.g., discriminant analysis [Murray and Horn 1989; Keough and Quinn 1991]), three types of analysis are probably most often used in studies of benthic marine communities. These are cluster analysis, nonmetric multidimensional scaling (MDS), and principal components analysis (PCA). All of these procedures use species-abundance data to calculate similarities between pairs of samples. Cluster analysis uses an algorithm to build a dendrogram displaying similarities between samples and groups of samples. MDS is an ordination technique that uses rank distances to represent similarities between samples in two or more dimensions of unitless space. MDS has proven to be a particularly valuable tool for marine ecological studies because it does not require the usual assumptions of multivariate normality and, therefore, readily works on zero-rich abundance data sets. MDS also can be used effectively on parameters such as variance-to-mean ratio (Warwick and Clarke 1993) or even simple presence/absence data (Prentice 1977; Field et al. 1982; Minchin 1987; Clarke 1993). Hence, if observers have the necessary taxonomic expertise, MDS of presence/absence or ranks data obtained from rapid surveys may prove to be quite powerful for detecting changes in whole communities. PCA, another ordination procedure, also produces plots of samples in two or more dimensions, with the axes serving more or less as lines of best fit so as to sequentially represent as much of the variation as possible in the data set. Problems are encountered with zero-rich species data, but PCA is particularly useful when applied to environmental data expressed in different measurement units. Also, PCA axes have units and it is possible to calculate statistical correlations between axes and original variables and to correlate PCA scores for sample groups, such as sites, to species abundance or other data. For example, cluster analysis was used by Littler and Murray (1975) to identify intertidal zones and communities impacted by sewage discharge (fig. 3.3). Warwick and Clarke (1991) and Gray et al. (1990) have used MDS to discriminate between polluted and unpolluted soft sediment sites and Schoch (1999) found that MDS ordination can discriminate effectively between communities found on rocky shores of different slope angles (fig. 3.4). Among other uses, PCA was selected to identify seasonal signals in environmental conditions in a study of cold-temperate–zone rocky intertidal macrophytes (Murray and Horn 1989).

PRIMER (Clarke and Gorley 2001) is a user-friendly Windows program that readily incorporates data from Microsoft Excel and other formats. PRIMER performs cluster, MDS, PCA, and several other informative analyses, including two other valuable procedures, ANOSIM (analysis of similarities [Clarke and Green 1988]) and SIMPER (similarity

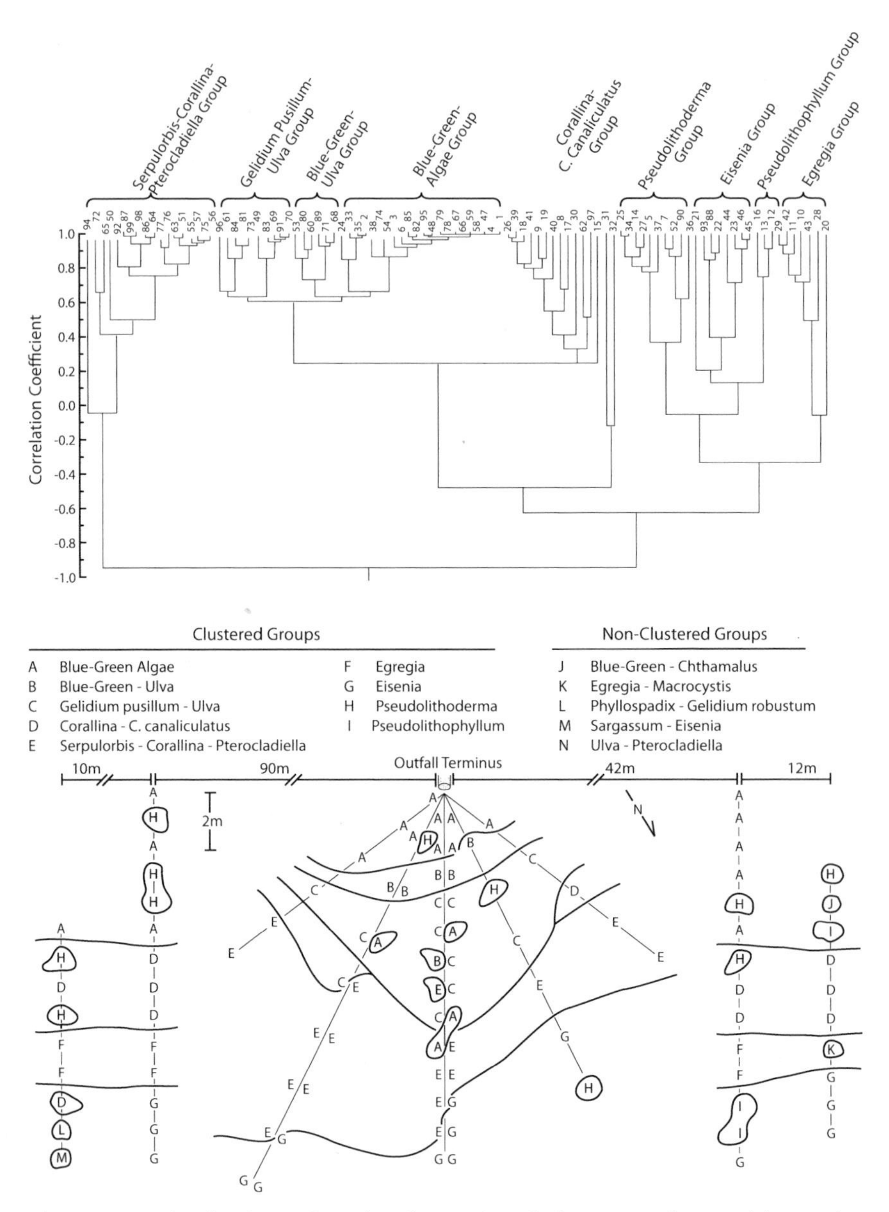

Figure 3.3. Distributions of species clusters in relation to quadrat position and locations of outfall and control transects at Wilson Cove, San Clemente Island, California, showing effects of sewage on rocky intertidal community structure. (Redrawn from Littler and Murray 1975 with updated nomenclature.)

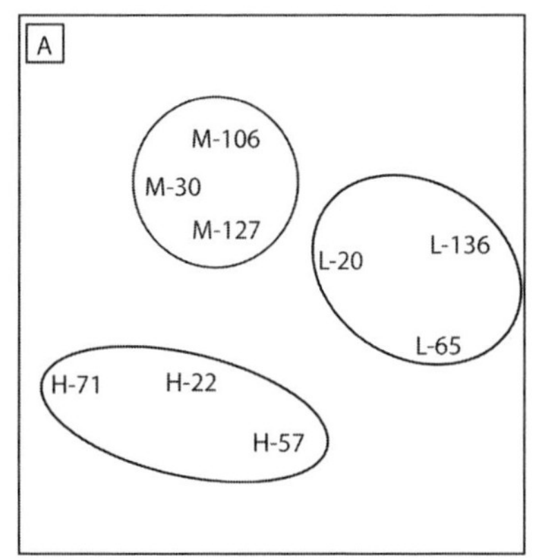

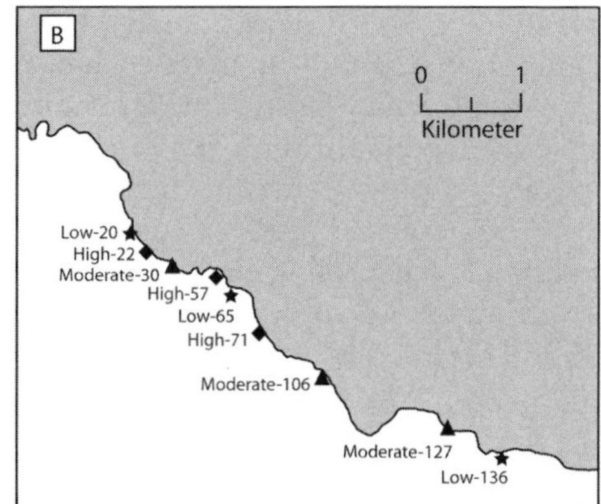

Figure 3.4. MDS comparisons of rocky shore community structure among three lower-zone segment groups on San Juan Island, Washington. MDS plot shows that beach segments are tightly grouped according to their respective slope groups and that slope groups are clearly separated in MDS space. MDS ordinations based on community data collected from beach segments with high (> 60°), moderate (20–35°) and low (<10°) slope angles (stress = 0.11). One-way ANOSIM with 280 permutations, global R = 0.87; significance level = 0.004 (pairwise tests with 10 permutations each: high vs. moderate, high vs. low, and moderate vs. low yielded significance levels of 0.1 for all comparisons). (Modified from Schoch 1999.)

percentages). ANOSIM, a procedure introduced to detect environmental impacts, is becoming widely used (e.g., Gray et al. 1988; Warwick 1988b; Chapman et al. 1995) for testing hypotheses about spatial and temporal changes in marine communities (e.g., assessing perturbed vs. not perturbed study sites). SIMPER provides the breakdown of the contribution of each species to similarity values calculated within and among sites or sample groups. Further discussion of these procedures can be found in Clarke (1993).

A disadvantage of many multivariate analyses is that they are better at identifying patterns than explaining them. This is because most are not designed for hypothesis testing; in addition, unlike factorial ANOVAs, they cannot test for interactions that may provide critically important information (Underwood 1996). For example, BACI analyses specifically seek interactions between locations (perturbed vs. unperturbed) and time (before vs. after) (Underwood 1994). However, multivariate procedures such as cluster analysis, MDS, and PCA can be used to identify patterns and then studies can be designed to test specific hypotheses on target species (e.g., those highlighted as important by the SIMPER procedure). Differences among MDS clusters also can be tested statistically using multiple response permutation procedures (Zimmerman et al. 1985). Philippi et al. (1998) note that while differences in species composition among areas may be striking, it is more difficult to define (or statistically analyze) trends through time in composition. In addition, changes in species composition may be caused by factors (e.g., succession) other than the environmental factor of concern. Overall, however, multivariate procedures are becoming increasingly useful tools for examining patterns involving most or all community taxa and hold promise for detecting subtle and ecologically important changes in community features.

SUMMARY

Of the biological units targeted in intertidal monitoring, impact-detection, and other research programs, the species-level (population) is the most commonly sampled unit. A major problem with species-level sampling is the high variability in abundances that occur through space and time; in addition, there is no *a priori* way to choose the "best" species to sample. Biomonitors or indicator species have yet to be defined for the range of stressors and changing environmental conditions that affect rocky intertidal communities. Species commonly followed during monitoring and other study programs are large, abundant, space-occupying taxa. If one or only a few species are to be studied, keystone species known to play an important role in organizing the community should be targeted. Politically expedient species, such as those of high charismatic value, or "millstone" species, ones that have been studied historically but without consideration of alternatives, should not be selected without ecological justification.

Down-a-level individual-based parameters such as gonadal production, growth rate, and size may be more sensitive than population parameters to unusually good or bad environmental conditions. The statistical power to detect change in these parameters also may exceed the power available using population-based abundance data such as density, because the latter are usually more variable among sample replicates. However, data for individual-based parameters are time-consuming to gather, and background data are needed to determine which species and parameters should be sampled.

Biological units consisting of lumped categories, either taxonomic (e.g., families) or functional (e.g., functional groups or guilds), can be sampled with greater efficiency and in some cases yield the sensitivity needed to detect community responses to changing environmental conditions and the presence of stressors. However, their use must be tested in each system or community before such groupings become the foundation of any intertidal sampling program. Community-level metrics such as species diversity provide information that integrates the whole suite of species in the community. Unfortunately, the sensitivity of various diversity measures to stressors or other environmental conditions has seldom been tested on rocky shores. This is a critical area in need of more research. One compromise approach for determining the biological units in an intertidal sampling program is to perform both population- and individual-based sampling on dominant taxa, together with qualitative sampling (e.g., presence/absence data) of all taxa in the community. In this way, multivariate tools can be used effectively to detect and analyze differences in community patterns. Multivariate tools such as MDS and ANOSIM are increasingly being used to successfully detect changing community patterns in marine communities and to relate these patterns to changes in environmental conditions.

LITERATURE CITED

Ambrose, R.F., P.T. Raimondi, J.M. Engle, and J.M. Altstatt. 1994. Inventory of shoreline resources in Santa Barbara County interim report. Coastal Research Center, Marine Science Institute, University of California, Santa Barbara.

Brosnan, D.M., and L.I. Crumrine. 1994. Effects of human trampling on marine rocky shore communities. *J. Exp. Mar. Biol. Ecol.* 177:79–97.

Cao, Y., and D.D. Williams. 1999. Rare species are important in bioassessment (reply to the comment by Marchant). *Limnol. Oceanogr.* 44:1841–42.

Cao, Y., D.D. Williams, and N.E. Williams. 1998. How important are rare species in aquatic community ecology and bioassessment. *Limnol. Oceanogr.* 43:1403–9.

Chapman, M.G., A.J. Underwood, and G.A. Skelleter. 1995. Variability at different spatial scales between a subtidal assemblage exposed to the discharge of sewage and two control assemblages. *J. Exp. Mar. Biol. Ecol.* 189:103–22.

Clarke, K.R. 1993. Non-parametric multivariate analyses of changes in community structure. *Aust. J. Ecol.* 18:117–43.

———. 1997. *Marine pollution.* 4th ed. Oxford, UK: Clarendon Press.

Clarke, K.R., and R.N. Gorley. 2001. *PRIMER (Plymouth Routines in Multivariate Ecological Research) v5. User manual/tutorial.* Plymouth UK: PRIMER-E Ltd., Plymouth Marine Laboratory.

Clarke, K.R., and R.H. Green. 1988. Statistical design and analysis for a 'biological effects' study. *Mar. Ecol. Progr. Ser.* 46:213–26.

Dethier, M.N. 1991. The effects of an oil spill and freeze event on intertidal community structure in Washington. OCS Study MMS 91–0002.

Engle, J.M., J.M. Altstatt, P.T. Raimondi, and R.F. Ambrose. 1994. Rocky intertidal monitoring handbook for Inventory of intertidal resources in Santa Barbara County. Coastal Research Center, Marine Science Institute, University of California, Santa Barbara.

Faith, D.P., C.L. Humphrey, and P.L. Dostine. 1991. Statistical power and BACI designs in biological monitoring: comparative evaluation of measures of community dissimilarity based on benthic macroinvertebrate communities in

Rockhole Mine Creek, Northern Territory, Australia. *Aust. J. Mar. Freshwater Res.* 42:589–602.

Farnsworth, E.J., and A.M. Ellison. 1996. Scale-dependent spatial and temporal variability in biogeography of mangrove root epibiont communities. *Ecol. Monogr.* 66:45–66.

Ferraro, S.P., and F.A. Cole. 1992. Taxonomic level sufficient for assessing a moderate impact on macrobenthic communities in Puget Sound, Washington, USA. *Can. J. Fish. Aquat. Sci.* 49:1184–88.

Field, J.G., K.R. Clarke, and R.M. Warwick. 1982. A practical strategy for analysing multispecies distribution patterns. *Mar. Ecol. Progr. Ser.* 8:37–52.

Fletcher, H., and C.L.J. Frid. 1996. Impact and management of visitor pressure on rocky intertidal algal communities. *Aquat. Conserv. Mar. Freshwater Ecosyst.* 6:287–97.

Gray, J.S., M. Aschan, M.R. Carr, K.R. Clarke, R.H. Green, T.H. Pearson, R. Rosenberg, and R.M. Warwick. 1988. Analysis of community attributes of the benthic macrofauna of Frierfjord/Langesundfjord and in a mesocosm experiment. *Mar. Ecol. Progr. Ser.* 66:285–99.

Gray, J.S., K.R. Clarke, R.M. Warwick, and G. Hobbs. 1990. Detection of the initial effects of pollution on marine benthos: an example from the Ekofisk and Eldfisk oil fields, North Sea. *Mar. Ecol. Progr. Ser.* 46:171–80.

Hay, M.E. 1994. Species as 'noise' in community ecology: Do seaweeds block our view of the kelp forest? *TREE* 9:414–16.

Heip, C., B.F. Keegan, and J.R. Lewis, eds. 1987. *Long-term changes in coastal benthic communities.* Dordrecht, the Netherlands: W. Junk.

Hixon, M.A., and W.N. Brostoff. 1996. Succession and herbivory: effects of differential fish grazing on Hawaiian coral-reef algae. *Ecol. Monogr.* 66:67–90.

Houghton, J.P., R.H. Gilmour, D.C. Lees, W.B. Driskell, S.C. Lindstrom, and A. Mearns. 1997. Prince William Sound intertidal biota seven years later—Has it recovered? 1997 International Oil Spill Conference, Paper 260, 679–86.

James, R.J., M.P. Lincoln Smith, and P.G. Fairweather. 1995. Sieve mesh-size and taxonomic resolution needed to describe natural spatial variation of marine macrofauna. *Mar. Ecol. Progr. Ser.* 118:187–98.

Karr, J.R. 1991. Biological integrity: a long-neglected aspect of water resource management. *Ecol. Appl.* 1:66–84.

———. 1994. Biological monitoring: challenges for the future. In Loeb and Spacie, 1994, 357–73.

Keough, M.J., and G.P. Quinn. 1991. Causality and the choice of measurements for detecting human impacts in marine environments. *Aust. J. Mar. Freshwater Res.* 42:539–54.

Kratz, T.K., J.J. Magnuson, T.M. Frost, B.J. Benson, and S.R. Carpenter. 1994. Landscape position, scaling, and the spatial and temporal variability of ecological parameters: considerations for biological monitoring. In Loeb and Spacie, 1994, 217–31.

Legendre, L., and P. Legendre. 1983. *Numerical ecology.* New York: Elsevier Scientific.

Lewis, J.R. 1976. Long-term ecological surveillance: Practical realities in the rocky littoral. *Oceanogr. Mar. Biol. Annu. Rev.* 14:371–90.

———. 1982. The composition and functioning of benthic ecosystems in relation to the assessment of long-term effects of oil pollution. *Philos. Trans. R. Soc. Lond. Ser. B Biol. Sci.* 297:257–67.

Lewis, J.R., R.S. Bowman, and M.A. Kendall. 1982. Some geographical components of population dynamics: Possibilities and realities in some littoral species. *Neth. J. Sea Res.* 16:18–28.

Littler, M.M., and K.E. Arnold. 1982. Primary productivity of marine macroalgal functional-form groups from southwestern North America. *J. Phycol.* 18:307–11.

Littler, M.M., and D.S. Littler. 1980. The evolution of thallus form and survival strategies in benthic marine macroalgae: field and laboratory tests of a functional-form model. *Am. Nat.* 116:25–44.

———. 1984. Relationships between macroalgal functional form groups and substrata stability in a subtropical rocky-intertidal system. *J. Exp. Mar. Biol. Ecol.* 74:13–34.

Littler, M.M., and S.N. Murray. 1975. Impact of sewage on the distribution, abundance and community structure of rocky intertidal macro-organisms. *Mar. Biol.* 30:277–91.

Loeb, S.L., and A. Spacie, eds. 1994. *Biological monitoring of aquatic systems*. Boca Raton, FL: Lewis.

López Gappa, J.J., A. Tablado, and N.H. Magaldi. 1990. Influence of sewage pollution on a rocky intertidal community dominated by the mytilid Brachidontes rodriguezi. *Mar. Ecol.* Progr. Ser. 63:163–75.

López Gappa, J.J., A. Tablado, and N.H. Magaldi. 1993. Seasonal changes in an intertidal community affected by sewage pollution. *Environ. Poll.* 82:157–65.

Magurran, A.E. 1988. *Ecological diversity and its measurement*. Princeton, NJ: Princeton Univ. Press.

Marchant, R. 1999. How important are rare species in aquatic community ecology and bioassessment? A comment on the conclusions of Cao et al. *Limnol. Oceanogr.* 44:1840–41.

Menge, B.A., L.R. Ashkenas, and A. Matson. 1983. Use of artificial holes in studying community development in cryptic marine habitats in a tropical rocky intertidal region. *Mar. Biol.* 77:129–42.

Minchin, P.R. 1987. An evaluation of the relative robustness of techniques for ecological ordination. *Vegetatio* 69:89–107.

Morrisey, D.J., L. Howitt, A.J. Underwood, and J.S. Stark. 1992. Spatial variation in soft-sediment benthos. *Mar. Ecol. Progr. Ser.* 81:197–204.

Murray, S.N., and M.H. Horn. 1989. Seasonal dynamics of macrophyte populations from an eastern North Pacific rocky-intertidal habitat. *Bot. Mar.* 32:457–73.

Murray, S.N., and M.M. Littler. 1984. Analysis of seaweed communities in a disturbed rocky intertidal environment near Whites Point, Los Angeles, California, USA. Hydrobiologia 116/117:374–82.

Oliver, I., and A.J. Beattie. 1996. Designing a cost-effective invertebrate survey: a test of methods for rapid assessment of biodiversity. *Ecol. Appl.* 6:594–607.

Olsgard, F., P.J. Somerfield, and M.R. Carr. 1997. Relationships between taxonomic resolution and data transformations in analyses of a macrobenthic community along an established pollution gradient. *Mar. Ecol. Progr. Ser.* 149:173–81.

Osenberg, C.W., R.J. Schmitt, S.J. Holbrook, K.E. Abu-Saba, and A.R. Flegal. 1994. Detection of environmental impacts: natural variability, effect size, and power analysis. *Ecol. Appl.* 4:16–30.

Paine, R.T. 1966. Food web complexity and species diversity. *Am. Nat.* 100:65–75.

Philippi, T.E., P.M. Dixon, and B.E. Taylor. 1998. Detecting trends in species composition. *Ecol. Appl.* 8:300–308.

Phillips, J.C., G.A. Kendrick, and P.S. Lavery. 1997. A test of a functional group approach to detecting shifts in macroalgal communities along a disturbance gradient. *Mar. Ecol. Progr. Ser.* 153:125–38.

Pielou, E.C. 1975. *Ecological diversity*. New York: Wiley–Interscience, John Wiley and Sons.

Posey, M., C. Powell, L. Cahoon, and D. Lindquist. 1995. Topdown vs. bottom up control of benthic community composition on an intertidal tideflat. *J. Exp. Mar. Biol. Ecol.* 185:19–31.

Povey, A., and M.J. Keough. 1991. Effects of trampling on plant and animal populations on rocky shores. *Oikos* 61:355–68.

Prentice, I.C. 1977. Non-metric ordination methods in ecology. *J. Ecol.* 65:85–94.

Schoch, G.C. 1999. *Untangling the complexity of nearshore ecosystems: examining issues of scaling and variability in benthic communities*. Ph.D. dissertation. Corvallis: Oregon State University.

Schroeter, S.C., J.D. Dixon, J. Kastendiek, R.O. Smith, and J.R. Bence. 1993. Detecting the ecological effects of environmental impacts: a case study of kelp forest invertebrates. *Ecol. Appl.* 3:331–50.

Sousa, W.P. 1979a. Disturbance in marine intertidal boulder fields: the nonequilibrium maintenance of species diversity. *Ecology* 60:1225–39.

———. 1979b. Experimental investigations of disturbance and ecological succession in a rocky intertidal algal community. *Ecol. Monogr.* 49:227–54.

Steneck, R.S., and M.N. Dethier. 1994. A functional group approach to the structure of algal-dominated communities. *Oikos* 69:476–97.

Underwood, A.J. 1994. On beyond BACI: sampling designs that might reliably detect environmental disturbances. Ecol. Appl. 4:3–15.

———. 1996. Detection, interpretation, prediction and management of environmental disturbances: some roles for experimental marine ecology. *J. Exp. Mar. Biol. Ecol.* 200:1–27.

Underwood, A.J., and C.H. Peterson. 1988. Towards an ecological framework for understanding pollution. *Mar. Ecol. Progr. Ser.* 46:227–34.

Underwood, A.J., and P.S. Petraitis. 1993. Structure of intertidal assemblages in different locations: how can local processes be compared? In *Species diversity in ecological communities,* ed. R.W. Ricklefs and D. Schluter, 39–51. Chicago: Univ. of Chicago Press.

Warwick, R.M. 1988a. The level of taxonomic discrimination required to detect pollution effects on marine benthic communities. *Mar. Pollut. Bull.* 19:259–68.

———. 1988b. Analysis of community attributes of the macrobenthos of Frierfjord/Langesundsfjord at taxonomic levels higher than species. *Mar. Ecol. Progr. Ser.* 46:167–70.

Warwick, R.M., and K.R. Clarke. 1991. A comparison of some methods for analysing changes in benthic community structure. *J. Mar. Biol. Assoc.* UK 71:225–44.

———. 1993. Increased variability as a symptom of stress in marine communities. *J. Exp. Mar. Biol. Ecol.* 172:215–26.

Weinberg, J.R. 1984. Interactions between functional groups in soft-substrata: do species differences matter? *J. Exp. Mar. Biol. Ecol.* 80:11–28.

Weisberg, S.B., J.A. Ranasinghe, D.M. Dauer, L.C. Schaffner, R.J. Diaz, and J.B. Frithsen. 1997. An estuarine benthic index of biotic integrity (B-IBI) for Chesapeake Bay. *Estuaries* 20:149–58.

Wilson, J.B. 1999. Guilds, functional types and ecological groups. *Oikos* 86:507–22.

Wilson, W.H. 1991. Competition and predation in marine soft-sediment communities. *Annu. Rev. Ecol. Syst.* 21:221–41.

Zeh, J.E., J.P. Houghton, and D.C. Lees. 1981. Evaluation of existing marine intertidal and shallow subtidal biologic data. Prepared for the MESA Puget Sound Project, Seattle, Washington. EPA Interagency Agreement No. D6-E693-EN.

Zimmerman, G.M., H. Goetz, and P.W. Mielke, Jr. 1985. Use of an improved statistical method for group comparison to study effects of prairie fire. *Ecology* 66:606–11.

Rockweed communities on a flattened, intertidal platform at Shaw's Cove, Laguna Beach, California.

CHAPTER 4

Sampling Design

In rare cases, it is possible to determine the abundance of an organism by counting all of the individuals in an area, making it simple to look for changes in a population or differences between populations. This is usually impossible, though, and so *sampling* is performed to provide an estimate of abundances or other parameters of interest. Regardless of the details of sampling methodology, the overarching goal is to obtain an accurate, unbiased estimate of a parameter. In addition the efficiency of a selected sampling approach is of concern, because the time or money available to carry out the study is always limiting.

This chapter focuses on decisions about the overall design of the sampling program, especially how many samples need to be taken and how the sampling units should be arranged at a site (fig. 4.1). Later chapters consider the types of sampling units (chapter 5), the particular methodology that can be used to assess individual parameters (chapter 8), and the abundances of populations (chapters 6 and 7).

The decision about what sampling design to use in a study is a fundamental decision that will determine what inferences can be made from data collected, and even whether the data will be valid for use in the study. The issues are primarily statistical, so this chapter begins with a brief discussion of statistical considerations. The next section covers the topic of how sampling units are positioned in a study area. Random, systematic, and targeted placement and the various methods for stratifying sampling units in the field are then discussed. The final section of this chapter briefly discusses how to determine the appropriate number of samples to take.

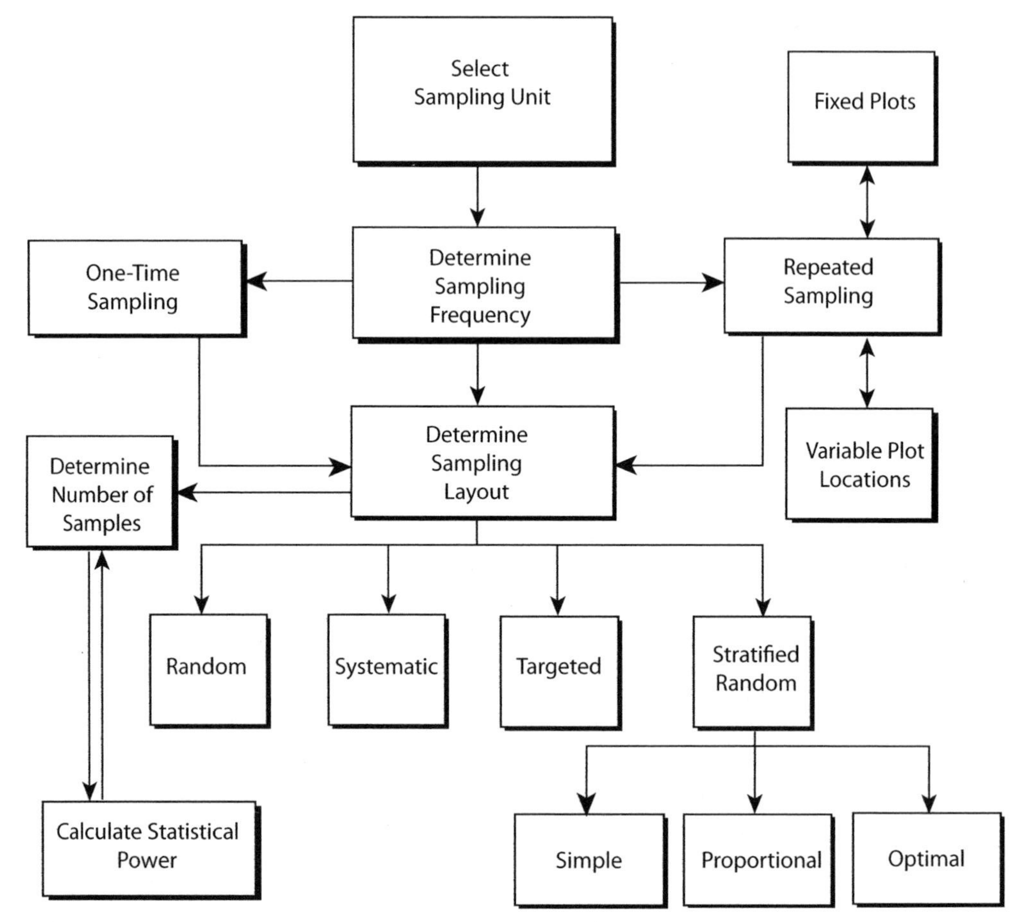

Figure 4.1. Decision tree for developing a sampling design.

STATISTICAL CONSIDERATIONS

Assumptions

To ensure that estimates are unbiased, and to satisfy the assumptions of most parametric statistical tests, the collected data must have independent and normal error distributions, homogeneity of error variation among groups, and additivity of effects (Green 1979). Independence of errors is perhaps the most critical assumption (Glass et al. 1972). As Green (1979) points out, it is the only assumption in most statistical methods for which "violation is both serious and impossible to cure after the data have been collected." The sampling design chosen will determine whether or not this assumption is met.

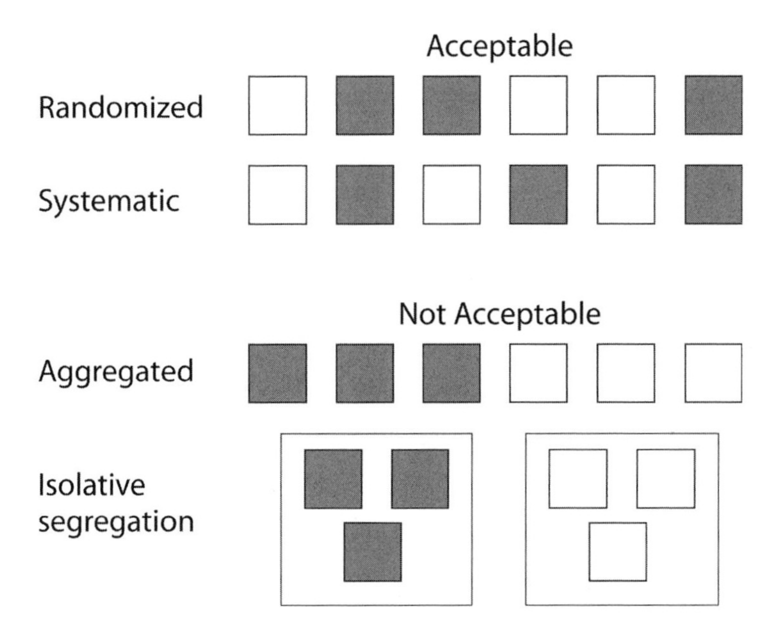

Figure 4.2. Examples of acceptable and unacceptable sampling designs. The rectangles represent sampling units, with shading representing different treatments. Here, unacceptable designs provide pseudoreplication. More examples and details are given by Hurlbert (1984).

Random sampling is generally considered necessary to ensure independence of errors. Individual samples need to be random to be considered replicate samples. A related problem concerning the lack of independence among replicate samples is the problem of "pseudoreplication." Pseudoreplication, which can take many forms, has been a problem in many ecological studies (Hurlbert 1984). A common cause of pseudoreplication in field studies is the incomplete randomization of sampling or experimental units. For example, in figure 4.2 the sampling schemes with quadrats clumped together are pseudoreplicates because these quadrats are not likely to be independent of each other, and hence, they are not true replicates. If the gray rectangles are control plots and the white rectangles are, for example, predator removal plots, then differences between treatments in the unacceptable designs might simply be due to natural differences on different parts of a rocky reef for the aggregated design or on different rocky reefs for the isolative segregation design.

TABLE 4.1. Alternatives in Hypothesis Testing and Relationships with Type I and Type II Errors

	Reality	
Decision	*H_0 True*	*H_0 False*
Reject H_0	Type I error (α)	Correct decision Power: $1-\beta$
Do not reject H_0	Correct decision	Type II error (β)

Violations of the other assumptions are common in ecological data. For example, abundances of organisms tend to be skewed, with a few abundant species and many rare species, and heterogeneity of variances often occurs when the variance depends on the mean. When data initially fail these assumptions, various transformations can sometimes be used to bring the data into conformity with the assumptions. Also, moderate failure of some assumptions, such as normality, has little effect on the outcome of statistical tests (Stewart-Oaten 1995). Glass et al. (1972) also discuss the practical consequences of violations of assumptions.

Statistical Power

In most cases, collected data will be used to determine whether there is a significant difference between different samples. There is often too much emphasis on making a Type I error when conducting statistical tests, that is, rejecting the null hypothesis when in fact it is true (table 4.1). In many uses of statistical tests, such as assessing the efficacy of a new drug, it is clear why a Type I error (i.e., deciding that the new drug is effective when in reality it is not) should be avoided. However, it is also important to consider the probability of making a Type II error, that is, failing to reject the null hypothesis when in fact it is false. Several researchers have pointed out that a Type II error may be especially serious when evaluating possible environmental impacts (Schroeter et al. 1993; Osenberg et al. 1994; Wiens and Parker 1995). In these cases, a researcher might mistakenly conclude that there is no environmental impact (i.e., fail to reject the null hypothesis) when in reality there is an impact. For this reason, an increasing number of researchers have emphasized the importance of considering the statistical power of an analysis (Andrew and Mapstone 1987; Peterman 1990; Fairweather 1991; Mapstone 1995).

Statistical power is defined as the probability that a test will lead to the correct rejection of the null hypothesis. Power is the complement of Type II error (β) and, so, is expressed as $1 - \beta$. Five parameters are relevant for analyses of power: power, significance level (α), variance, sample size, and effect size. Knowing (or specifying) any four of these allows calculation of the fifth for a particular statistical procedure. Power increases with increasing effect size and higher α level and decreases with increasing sample variance. Because the variance of a sample decreases with increasing sample size, a change in sample size influences power. For this reason, power analysis is often used to determine the sample size needed in a study.

As discussed above, there is an implicit trade-off between Type I and Type II error. The relationship between power and Type I error (α) is not a simple one. Some of the relationships of most interest for monitoring programs include the following. If sample size is held constant, as is often the case for monitoring programs because of time or money constraints, α and power will be directly related; for example, relaxing α from 0.05 to 0.10 will result in increased power to detect an effect size or difference. If effect size and α are held constant, then power will increase as sample size increases.

A discussion of the actual steps required for the calculation of power is beyond the scope of this chapter. Traditionally, few ecologists have calculated the power of their statistical tests, perhaps because it was not easy or convenient to do so. It is still possible to calculate power using tables or charts provided by many sources (e.g., Cohen 1988). However, there now exist a wide variety of computer programs for calculating power, some of which are very user-friendly. Thomas and Krebs (1997) reviewed many of these programs.

Power cuts across many levels of sampling design. A common cause of low power in ecological studies is the high degree of spatial and temporal variability that is characteristic of most communities (Osenberg et al. 1994), in particular, rocky intertidal communities. Thus, many of the issues concerning selection of sampling units and the layout and disposition of sampling units revolve around means of reducing unnecessary variation ("noise") and obtaining sufficient numbers of samples to achieve the level of statistical power required to detect the effect size or "signal" of interest.

Finally, the proper interpretation of statistical tests requires the researcher to distinguish between statistical significance and ecological significance. Statistics tell us how likely it is, for example, that a difference in the abundance of a species from one time period to the next is due to chance alone. Statistics do not tell us whether such a difference is important from an ecological perspective. In some cases, efficient sampling designs or large sample sizes can give high power to detect very small differences, differences that may not matter in the context of the ecological processes occurring in the system. The use of statistics is important for determining whether a difference is real, but it is up to the researcher to state whether the detected difference is of ecological importance. There are no simple guidelines for determining what effect size matters. However, given the wide fluctuations in abundances experienced by many marine algae, invertebrates, and fish, the goal of detecting a 50% change in abundance with 80% power has frequently been adopted for environmental assessments.

LOCATION OF SAMPLING UNITS

One of the critical decisions in designing a sampling program is how to place the sampling units in the study area. This decision determines the nature of the information collected and, thus, the accuracy and the inferences that can be drawn from these data. Many books and papers discuss this topic. For example, Cochran (1977) provides a general discussion, Green (1979) directs his overview at environmental biologists, and Greig-Smith (1983) focuses on vegetation sampling (which, because of the sedentary nature of the target organisms and a frequent focus on cover, addresses many of the same issues encountered in intertidal sampling). Andrew and Mapstone (1987) and Kingsford and Battershill (1998) discuss sampling design issues related to general marine ecology, while Gonor and Kemp (1978) focus specifically on intertidal sampling.

Distribution of Sampling Effort

The general goal of sampling in a particular area is to obtain representative samples of the population(s) of interest. Several different approaches can be taken, as discussed in the following sections. However, even before the approach for locating sampling units is determined, a basic question about the sampling program must be answered: What will and will not be studied? The answer to this question influences both how the samples should be located in the study area and what sorts of inferences can be made from the data; for this reason, the decision should be made before the study begins. For example, in rocky intertidal monitoring or impact studies one often must decide whether or not tidepools should be sampled. If tidepools are to be excluded, then sampling units that would ordinarily be placed in tidepools are not sampled and, instead, are relocated elsewhere. Operationally, some rules about excluding some sampling locations need to be decided beforehand. For example, tidepools need to be defined (e.g., persistent water >5 cm deep), and rules for rejecting a particular location (e.g., reject if >40% of the quadrat is covered by a tidepool) and for relocating the sampling unit (e.g., reassign random coordinates or flip quadrat upcoast until a valid location is encountered) need to be decided. Finally, the decision to exclude tidepools influences the inferences that can be drawn from the data collected, since the data would then relate to nontidepool rocky intertidal habitats only.

Establishing Sample Locations

Random. As discussed earlier, independence of errors is one of the most critical assumptions for statistical analyses of data, and random sampling is the best way to assure that this assumption is met. There are several different approaches to positioning randomly located sampling units, whether for a one-time survey or for repeated sampling when the quadrats are located randomly for each sampling period (fig. 4.3). The most straightforward approach is to lay two transect lines down at right angles to each other as axes for a coordinate system. The transects can be located along the periphery of the study site (as an "L") or through the center of the site (as a "+"). A pair of random numbers is chosen for each quadrat, and the corresponding position along the axes located. The most precise means of locating a particular position would be to use one or two transect lines to determine distances, but a rough measure, such as pacing, is sufficient and much faster and more convenient (and can be employed by one person).

A second approach to randomly locating sampling units is to use the "random walk." From an arbitrary starting point, a pair of random numbers is used to determine the distance and direction to the next sampling location. From that location, a new pair of random numbers is used to determine the distance and direction to the next sampling location, and so on. A variation on this approach is to use a number of fixed positions placed throughout the study area, and to use pairs of random numbers to determine distance and direction for sampling locations, each determined from the fixed position. After sampling one to several locations at one position, the process is repeated for each of the other fixed positions, so there is in effect a "circle" of sampling units, at varying distances and directions, around each fixed position. Although a GPS (global positioning system) could be used to locate quadrats randomly without having to resort to the above methods, portable units that supply sufficient accuracy are not yet economically available to meet the needs of most rocky intertidal sampling applications.

Note that haphazard sampling is not random sampling. Greig-Smith (1983) states that haphazard sampling of a plot, such as by walking over

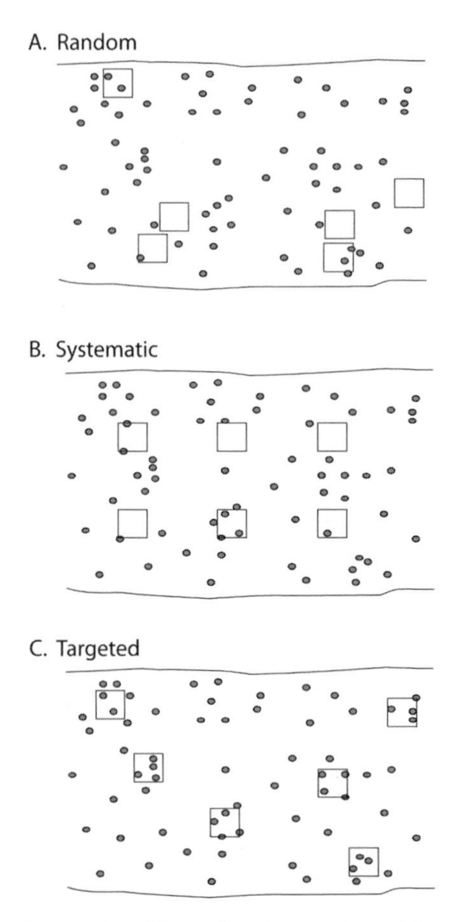

Figure 4.3. Schematic drawing illustrating (A) random, (B) systematic, and (C) targeted approaches to locating sampling units. Random quadrat locations were determined using a random number generator; targeted quadrat locations were determined "by eye." In this example, the true density is 8.3 individuals/m^2; random = 9.3, systematic = 11.1, and targeted = 33.3 individuals m^{-2}.

an area and throwing the quadrat over a shoulder, will almost always result in a nonrandom distribution of sample units. Sampling units are likely to be more spaced out than would occur with truly random placement, and the edges of the study area are likely to be underrepresented. Greig-Smith asserts that the extra trouble of employing a more objective method of randomizing, such as laying down two lines at right angles as axes and using a pair of random numbers as coordinates to position each sample unit, is worth the effort.

In spite of the advantages of randomization, many ecologists nonetheless locate their quadrats in a haphazard way. The approach suggested by Greig-Smith can be quite time-consuming to implement, especially over a large rocky intertidal area, and logistically difficult because of the presence of large tidepools, crevices, and other topographical features that characterize rocky intertidal habitats. In theory, haphazard sampling avoids the biases that might arise when an investigator places sampling units in "appropriate" locations. Of course, the possibility of conscious or subconscious bias still remains. Haphazard sampling trades off statistical rigor for convenience; whether this trade-off is worthwhile depends on the circumstances. For example, this trade-off may be acceptable for a quick informal survey of a site for an individual investigator, but not for samples taken for an impact assessment, which may later be subject to extreme scrutiny in a legal setting.

Systematic. One alternative to truly random location of sampling units is to place the sampling units systematically through the study area. A common approach to systematic sampling is to place sampling units uniformly across a study site, as in a grid. Greig-Smith (1983) discusses the advantages and consequences of systematic sampling. The two main advantages are as follows. (1) The estimate of the mean may be more accurate than with random samples. This is because the sampling units are spread throughout the study area, whereas in random sampling some areas may be sampled more intensely than others just by chance alone. (2) It may be easier to carry out systematic sampling than random sampling. Of course, the main disadvantage is that the samples are not randomly taken, and hence the basic statistical assumption of independence of errors might not be met. Greig-Smith (1983) states that there is no indication of the precision of the mean with systematic sampling and, hence, no possibility of assessing the significance of its difference to a mean obtained using similar procedures in another area.

One objection raised about systematic sampling is that it can produce misleading results if the spacing of sampling units corresponds to some underlying spatial pattern in the community. For example, sandy beaches often have cusps whose pattern is determined by the prevailing wave pattern, and the biota on the cusps may differ from the biota between cusps. If sampling units are uniformly placed on the same scale as the cusps, sampling may always occur on the cusps rather than between them, and the results will not reflect the actual beach community. Therefore, whenever systematic sampling is used, the investigator should carefully consider whether there might be some underlying pattern of natural variation that could bias the results.

Targeted. A third possible approach to locating sampling units is for the investigator to decide where to place each unit. For estimating parameters such as average density in a habitat, this approach is unacceptable (Greig-Smith 1983). Investigator bias in the choice of location is very likely with this approach, and hence the assumption of independence of errors may be severely violated. For most applications, a method of removing investigator choice in the location of sampling units (i.e., random, haphazard or systematic locations) is essential in order to be able to analyze the data statistically.

Although investigator bias in locating sampling units is to be avoided, there are of course many legitimate choices investigators make about where sampling units should be located. For example, particular habitat types or species assemblages may be targeted for study. When a subset of a study area is chosen for investigation, sampling outside the chosen area is unnecessary and inefficient. For example, if an investigator is interested in the species occurring in surfgrass beds, there is no point in sampling outside surfgrass beds. Restricting sampling to the area of interest is common sense. Within the targeted area, however, it is best if sampling units are located randomly without investigator bias.

Stratification. Since organisms are not distributed evenly throughout the intertidal zone, the variability associated with their cover can be markedly different over a small area. A varying spatial pattern within a study area can result in an overall reduction in sampling precision (Andrew and Mapstone 1987). When a study area is not homogeneous, *stratification*, the subdivision of an area into more homogeneous areas, with samples allocated among these subdivided areas, can be used to reduce the influence of spatial variability. If the chosen stratification is meaningful, the overall precision of the sampling effort will increase as a result of increased sampling precision within subdivisions. A good overview of stratification is given by Andrew and Mapstone (1987), with Cochran (1977) providing a more detailed, quantitative presentation of the topic.

Once stratification is imposed, a decision must still be made about how to allocate sampling units among the strata. Several methods can be used. In the simplest case (stratified simple random sampling [see Andrew and Mapstone 1987]), no information about the nature of the strata is used, and an equal number of samples is allocated to each stratum (fig. 4.4A). In a slightly more sophisticated approach, sampling units are allocated in proportion to area (proportional stratified sampling [Andrew and Mapstone 1987]). For example, if 100 sampling units were to be allocated to three strata covering 20%, 50%, and 30% of the site, then 20, 50, and 30 quadrats would be allocated, respectively. In both of these methods, no

A. Simple stratification

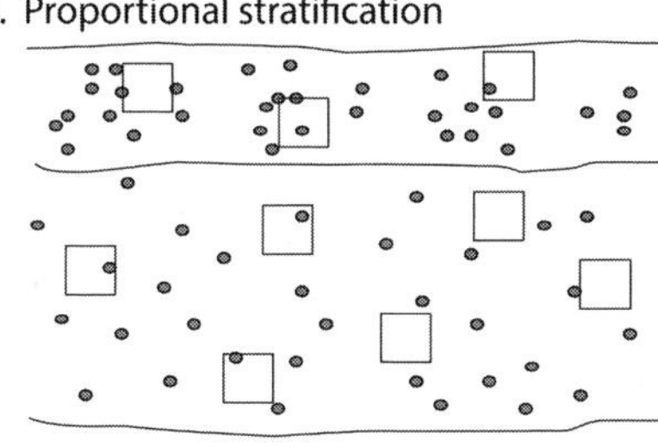

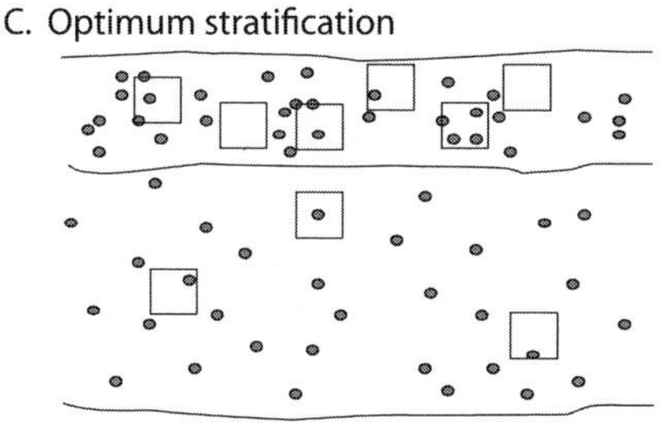

Figure 4.4. Location of sampling units using different methods of stratification. Shaded symbols represent the occurrence of individuals to be sampled. (A) Stratified simple random sampling. Equal numbers of sampling units are allocated to each stratum. (B) Proportional stratified sampling. Sampling units are allocated in proportion to the area of each stratum. In the example, the lower stratum has twice the area of the upper stratum, and so has six quadrats, compared to three quadrats in the upper stratum. (C) Stratified sampling with optimal allocation. Sampling units are allocated on the basis of within-quadrat spatial variance within each stratum. In the example, the upper stratum has twice the within-quadrat variance, and so has six quadrats, compared to three quadrats in the lower stratum.

information about differences in spatial distributions among the strata is used (fig. 4.4B).

The most sophisticated method allocates sampling units on the basis of per-quadrat spatial variance of each species within each stratum, with more samples allocated to areas with higher variances, in a so-called "optimum" allocation scheme (fig. 4.4C), (Cochran 1977; stratified sampling with optimal allocation [Andrew and Mapstone 1987]). By adjusting the number of samples to the variability within each stratum, the precision of the estimates for each stratum can be maximized. That is, few samples need to be taken in a stratum with very uniform distribution of a species in order to obtain a precise estimate of its mean abundance, whereas a stratum with high variability will require many samples in order to obtain an estimate of the mean with similar precision. (Cochran [1977] discusses several other criteria [in addition to minimizing the variance of the mean] for allocating quadrats according to the optimum allocation scheme.)

For an optimum allocation scheme, an estimate of variation is needed for each stratum before sampling can begin. Pilot studies can be used to obtain these estimates (Andrew and Mapstone 1987). On the other hand, simple visual estimates of species cover within each of the strata, which can be made without any actual sampling, provide an acceptable method of obtaining estimates of variation without having the additional time and expense of pilot studies (W. Cumberland, pers. comm.). Spatial variance estimates can be calculated as $v_{ij} = \text{sqrt}[(\text{cover}_{ij}) * (1 - \text{cover}_{ij})]$, where v_{ij} denotes the spatial variance of the ith species in the jth stratum (Miller and Ambrose 2000). The variance is calculated for each stratum, then normalized by dividing by the sum of variances for all strata. For example, if there are four strata, the normalized variance for species 1 in stratum 1 is $V_{1,1} = v_{1,1} / (v_{1,1} + v_{1,2} + v_{1,3} + v_{1,4})$. The sample size for species 1 in stratum 1 can be calculated by multiplying the total number of sample quadrats to be allocated across the entire site by $V_{1,1}$.

Regardless of the approach used, quadrats are positioned within each stratum just as they would be if the entire study area were treated as a single unit (e.g., randomly distributed). An overall estimate of a parameter such as species abundance is made by combining the estimates from individual strata, weighted appropriately.

Simple stratification schemes are fairly common in intertidal research. The most obvious stratification is along elevational zones, such as in the low, mid, and high intertidal zone. We are not aware of any intertidal studies employing an optimum allocation scheme. As noted below, Miller and Ambrose (2000) found that stratified random quadrats had better accuracy than purely random quadrats for five of six species examined, and that the optimum quadrat allocation method was usually more accurate than allocation in proportion to stratum area.

In practice, quadrats are commonly stratified along vertical transects (i.e., along elevational gradients) (Ambrose et al. 1995) rather than in the proportional or optimum allocation procedures. This is a logistically simpler way to approach stratified random sampling; rather than having to place quadrats a random distance along two dimensions, distance along only one dimension (along the transect) is varied randomly within each stratum. The cost of the simplified logistics is a greater potential for spatial autocorrelation, because the quadrats are constrained to be closer to one another than would be the case for truly random stratified sampling. It seems likely that the problem of greater spatial autocorrelation will be small, but in any case stratified random quadrats along transect lines will not be more accurate than purely stratified random quadrats, and so they will generally be less accurate than transects (Miller and Ambrose 2000).

COMPARISON OF APPROACHES

In spite of the clear importance of the method used for laying out sampling units, few studies have actually studied the implications of different approaches. Miller and Ambrose (2000) used computer-simulated subsampling of actual rocky intertidal data to compare different approaches for distributing the sampling effort across a study area. They compared the accuracy of random points, transects, randomly placed quadrats, proportional stratified quadrats, and optimal stratified quadrats (Miller and Ambrose 2000). Randomly placed, single point contacts provided the most accurate estimates of cover. With quadrats, some form of stratified random sampling usually gave better accuracy than simple random placement. In nearly all stratified cases, optimum allocation of sample units, where quadrats are allocated among strata according to the amount of variability within each stratum, yielded a higher accuracy than did allocation in proportion to the area of the strata. With one exception, line transects placed perpendicular to the elevational contours ("vertical transects") approached or exceeded the accuracy of the best-stratified quadrat efforts. The greater accuracy of line transects may be related to the maximum linear dimension of the sampling unit and the spatial patterns of the species studied. All six species investigated by Miller and Ambrose had aggregated distributions. In this situation, especially, sample points that are close to each other are more likely to yield similar values than points that are far apart (Palmer and White 1994; Palmer and van der Maarel 1995; Hurlbert 1984). Randomly placed quadrats may have been more likely to fall entirely within or entirely between patches than line transects were. The lower accuracy of random quadrats is due to the fact that they constitute groupings of points, so they capture too much spatial autocorrelation. Transects spread the points over a larger area and thus are less influenced by spatial autocorrelation than quadrats, but even the points along transects are not independent. With quadrats and transects, some level of pseudoreplication, and consequently inaccurate estimate of cover, is inevitable. Increasing the sampling effort improved the accuracy of the cover estimates, but for a given effort, transects provided a better estimate of cover than randomly placed quadrats.

The "low" sampling effort used by Miller and Ambrose (2000), equivalent to sampling a transect every 3 m, yielded an accuracy of roughly 50%. As an illustration of what this sort of accuracy means, consider the case of *Balanus*. The cover of *Balanus* at one of the study sites, White's Point, was 10.6%. If one were to sample using vertical transects every 3 m, the first sample could easily yield an estimate of 15% (the upper quartile of the simulated sampling) and the second sample an estimate of 5% (the lower quartile). If these samples were taken on different dates, one might conclude that *Balanus* cover had declined sharply, even though the actual cover had not changed. This modeled effort was a reasonable one for most field sampling. For example, recent baseline studies along the southern California coast used line transects spaced every 3 m along a 60-m baseline (Engle et al. 1995), the same spacing used for the low-effort simulations by Miller and Ambrose (2000). Even at higher sampling efforts, accuracy was such that one-quarter of the time, cover estimates for most common intertidal species can be expected to be more than 10%–15% above or below the true cover value due simply to within-site sampling error.

The accuracy of estimates for rare species was consistently poor since sampling units often missed such species altogether. Species richness was substantially underestimated by all sampling approaches tested. It appears that the traditional approaches to sampling intertidal organisms are insufficient for assessing the number of species at a site. Accurate estimates of species richness may require supplemental sampling specifically for that purpose.

These results illustrate the great influence that choice of sampling unit and location has on the accuracy of cover estimates in rocky intertidal communities. The importance of choosing an appropriate sampling method and placement is magnified when species have patchy distributions. If species were randomly or evenly dispersed, the different sampling approaches would make little difference. But the real distribution of rocky intertidal organisms is usually not random or uniform, so decisions about sampling units and placement can substantially influence the accuracy of survey results. Ecologists must be aware of these effects and design studies, experiments, and monitoring programs accordingly.

Even the best sampling design is likely to have limited accuracy in habitats that have as much spatial variability as the rocky intertidal zone.

One-quarter of the time, a reasonable sampling effort using the most accurate sampling unit and placement would likely yield estimates of the mean that are more than 25% higher or lower than the true mean. Tripling the sampling effort may improve the accuracy in estimating cover from ±25% to ±10%–15% for common species, but this probably represents a practical limit for accuracy. Even at this higher intensity, the common sampling approaches cannot accurately estimate the cover of rare species or the species richness of a site. For general surveys, a rough estimate of the cover of different common species may be sufficient for an overview of a rocky intertidal community. However, for many purposes, such as environmental monitoring, impact assessment, and biodiversity surveys, this level of accuracy is likely to be insufficient, and alternative approaches (e.g., fixed plots, visual scans) will be required.

One-Time ("One-Off") Versus Repeated Assessments

Two types of sampling can be planned for a site. A one-time or "one-off" assessment provides a snapshot of species abundances or distributions at a site, usually as a part of a larger program examining a number of different sites or for collecting data to examine a particular ecological question. In this type of assessment, an accurate estimate of abundance over the study area is likely to be a high priority. Any of the methods discussed above for locating sampling units could be used for such one-time assessments. Another approach is to conduct repeated assessments at a particular site at a number of different times. Although the latter type of sampling may be part of a larger program examining a number of different sites, the temporal component means that an accurate estimate of temporal change at each individual site will likely be an important study goal.

Two approaches to the layout of sampling units in repeated assessments are commonly used. The simplest approach is to locate sampling units exactly as if performing a one-time assessment. For example, if a stratified random quadrat approach is to be used, the same strata would be used for each sampling period, and new random locations within each stratum would be sampled each time. An alternative approach is to sample the exact same locations during each sampling period. Methods for laying out random locations have been discussed above; in the following section, issues that are unique to fixed plot sampling designs are discussed.

Fixed Plots

A fixed plot provides specific information about changes that occur in the area circumscribed by that plot over time. Because each sample is taken from the same location, differences between sampling periods can be confidently assigned to actual changes occurring in plot contents. In contrast, if new sampling locations are established during each sampling period, changes detected between sampling periods might alternatively be due to differences among locations in addition to differences over time. Fixed plots are used to provide better "signal" resolution by reducing "noise" resulting from samples being taken in different locations during different sampling periods.

The initial step for establishing fixed plots is no different from the approaches discussed above; that is, a random or stratified random design can be used, or particular areas or assemblages can be targeted. However, the decision on how to locate the sampling units has important consequences for data interpretation, and the rationale for positioning fixed plots might differ from the rationale used in a one-time assessment. Some sort of randomized location of sampling units will provide the best estimate of the abundances of species at the site, as well as the strongest statistical foundation. However, randomizing plot locations, even in the context of a stratified random sampling design, will still result in substantial variation among sampling units, and unless an extremely large sample size is used, it is possible that the high level of variation among samples will make it difficult to detect temporal differences in mean abundances (Dethier and Tear, pers. comm.). Recently, several monitoring programs have adopted a "target assemblage" approach, wherein sampling units are located in particular assemblages of intertidal species, rather than randomly located throughout the study site (Richards and Davis 1988; Ambrose et al. 1995). Consistent with the decision to use fixed plots to reduce variability in the data, targeting sampling locations on particular species assemblages further reduces variability. Long-term monitoring data from fixed plots located in target assemblages has a high power to detect changes in the abundances of common intertidal species (Spitzer, Ambrose, and Raimondi, unpubl. observ.).

In theory, even sampling units in targeted assemblages, such as barnacle and mussel communities, could be located in a random manner. However, the occurrence of target species in small, discontinuous patches and the need for sample location to match the dimensions of sample plots mean that randomization is often logistically difficult. Recent monitoring programs (Richards and Davis 1988; Ambrose et al. 1995) have used investigator choice to locate fixed plots. Plots were established at five locations using as criteria (1) the best cover or abundance of the species making up the targeted assemblage and (2) topographical features such as relatively flat horizontal rock surfaces of sufficient shape and size to facilitate photographic analysis. Because these plots are not random samples of the targeted assemblages, some statistical comparisons are not possible. Imagine that one location had an extensive area of high-density mussel beds, while another location had only a small area of high-density mussels plus scattered mussels elsewhere. If all five quadrats could fit into the high-density mussel assemblage at both sites, the mean cover would be similar even though the overall mussel abundance is clearly higher at the first site. Therefore, it is not legitimate to compare the cover of mussels at one site to the cover at another site. However, the main purpose of many monitoring programs is to look for changes in abundances over time, and the temporal dynamics of abundance data obtained from fixed plots can be analyzed statistically.

When fixed (permanent) plots are used, it is important to mark them so that they can be relocated for subsequent resampling. Simple approaches such as hammering in nails or placing small disks of marine epoxy at quadrat corners may suffice for short-term studies in well-known locations, but such marks will not persist or be easy enough to find when many sites are followed over an extended period of time. More permanent marking can be achieved by drilling holes into the substratum, inserting plastic screw anchors, and then screwing in stainless-steel or brass screws. These markers last well for a number of years but, after decades, can be overgrown by biota and become hard to locate (in part because of their small size). For very long-term monitoring, marine epoxy (e.g., Z-Spar) can be used to cement large stainless-steel bolts ($\frac{1}{4}$- to $\frac{3}{8}$-in. diameter) into holes drilled with battery- or gas-powered drills. These large bolts can be easily relocated (if human use in an area is low, 5–8 cm of bolt can be exposed), which can save an amazing amount of time during subsequent sampling, and will persist for decades. If the bolts are big enough, a sensitive metal detector can be used to relocate hidden bolts when the general area of the marker is known but the specific marker cannot be found. This is especially useful in mussel beds, where a thick cover of mussels can grow over bolts, making it impossible to locate them visually without severely disrupting the bed.

Regardless of the type of marker used, it is important to carefully map each quadrat in case markers are lost or covered (like in mussel beds) and difficult to locate. Reference bolts, large bolts placed strategically throughout the study area, are useful for mapping. Measurements (distance and bearing) to each quadrat from a minimum of two reference bolts should be taken, so that quadrats with missing markers can be relocated by triangulation. If an accurate GPS is available, readings can be made at each quadrat. This will greatly facilitate the construction of accurate site maps if the coordinates are input into a geographic information system (GIS). However, economical GPS systems currently are unable to take the place of accurate distance and bearing readings for relocating lost markers in the field.

It is also very useful to take photographs of the overall site and the area around each quadrat from known reference points. A number of photographs should be taken from different angles with all nearby quadrat markers in place. If it is especially likely that the markers will be hidden (as in a mussel bed, although we have also found that it can be surprisingly difficult to locate bolts even in low *Endocladia muricata* algal turfs), a large number of photos will prove to be useful. Photos can be placed in an album or sealed in plastic laminate and brought into the field, along with site maps and measurements, to facilitate plot relocation.

Comparison of Fixed Versus Nonfixed Plots

Although many of the advantages and disadvantages of using fixed versus nonfixed plots have been mentioned above, these issues are summarized here. Random plots give an unbiased estimate of what is at the site, which is what one wants for comparing species abundances at different sites. Because of the extreme spatial heterogeneity that characterizes most rocky intertidal sites, however, a large number of samples must be taken at each site to obtain reasonably accurate estimates of abundance and to obtain sufficient statistical power to detect ecologically meaningful abundance differences between or among sites. Also, if sampling locations are newly randomized during each sampling period, comparison of abundances from one time to the next can be compromised by the confounding of spatial and temporal changes; that is, changes from one sampling period to the next could be due to the fact that sampling locations were different between sampling periods. These problems can be minimized by taking enough samples, but sufficiently large sample sizes rarely can be taken in intertidal studies because of limitations on field time and the funds available for field sampling.

Fixed plots give up the ability to estimate overall abundances at a site in return for achieving lower variability in the measured abundance parameter (e.g., density or cover) among sampling units and greater statistical ability to detect changes over time. If sampling locations are originally determined at random, then the initial samples can be considered representative of the study area. However, subsequent samples may not represent the entire area. Moreover, the numbers of sampling units that might be satisfactory for other, less heterogeneous habitat types are likely to yield very high variability in the rocky intertidal zone because of the patchiness of species distributions (Dethier and Tear, pers. comm.). Regardless of how the plot locations are determined, population data from fixed plots must be analyzed using repeated-measures ANOVA models because the same organisms will be sampled repeatedly.

Targeted plot locations, as used by Richards and Davis (1988) and Ambrose et al. (1995), reduce the variability among plots and therefore provide better statistical power for detecting changes in abundances over time (Spitzer, Ambrose, and Raimondi, unpubl. observ.). However, these plots are not random samples of a population and, so, cannot be considered representative of the abundances of species at a site. Therefore, it is meaningless to compare abundances at one site to abundances at another when the sampling design incorporates fixed plots. Thus, inferences about abundances per se are limited in fixed plot sampling designs to within-site changes over time. However, it is possible to compare the *dynamics* of species abundances at different sites. For example, it is possible to determine whether abundances generally increased or decreased at the same time at a number of different sites.

NUMBER OF SAMPLES: HOW MANY IS ENOUGH?

Gonor and Kemp (1978) and Mace (1964) discuss some of the issues related to determining the number of samples that should be taken. Andrew and Mapstone (1987) also discuss this topic and argue for the value of pilot data for making this and other important decisions when designing a sampling program.

The appropriate number of samples needed for a particular study depends on a variety of factors, including the density and spatial distribution of the species, the level of precision desired, and the goal of the study. Andrew and Mapstone (1987) cite a number of papers reviewing studies concerning sample sizes, mainly for benthic organisms. One generalization emerging from these studies is that the number of samples needed to achieve a particular precision is closely related to population density (and the size of sampling units used): as the density of a population increases, the number of samples needed for a particular precision declines. Although these previous studies may provide a general starting point for determining the number of samples needed in a particular study, it is at best a rough starting point. Andrew and Mapstone (1987) argue that pilot studies are the best means of making these sorts of decisions.

Kingsford and Battershill (1998) present three approaches for determining the number of samples needed for a study. First, they propose plotting the number of replicates versus SE/mean (where SE = standard error = SD/$\sqrt{n}$), with data taken from pilot sampling. This will yield a negative decay curve, and the appropriate number of samples is determined by inspection (where the curve levels off, and there is little improvement in SE/mean with more replicates). Second, they present a formula from Andrew and Mapstone (1987) where a desired level of precision is required.

$$n = \left[\frac{SD}{p \cdot \bar{x}}\right]^2 \tag{4.1}$$

where p is the desired precision (e.g., 0.15) as a proportion of the mean, $\bar{x}$ is the mean of the samples, and SD is the standard deviation of the

TABLE 4.2. Hypothetical Data Showing the Abundance of Species *x* in 20 Quadrats

Quadrat No.	Abundance	SE/Mean
1	25	
2	15	0.250
3	20	0.144
4	22	0.102
5	29	0.106
6	30	0.098
7	26	0.083
8	20	0.076
9	19	0.072
10	16	0.073
11	27	0.067
12	10	0.081
13	39	0.091
14	25	0.084
15	27	0.078
16	21	0.074
17	27	0.069
18	19	0.067
19	26	0.063
20	23	0.060
Mean	23.3	
SD	6.28	

NOTE: SE/mean gives the SE as a proportion of the mean for the previous quadrats; for example, SE/mean for quadrat 5 is calculated based on values in quadrats 1 through 5.

samples. Finally, they present a formula from Snedecor and Cochran (1980, 441) for the number of samples required to obtain an abundance estimate with an allowable error, in terms of confidence limits.

$$n = \frac{4s^2}{L^2} \qquad 4.2$$

where L is the predetermined allowable error (size of 95% confidence limits) on the sample mean and s^2 is the sample variance.

These three approaches are illustrated with hypothetical data in table 4.2. The plot of sample size versus SE as a proportion of the mean shows a declining function, where initially the addition of more quadrats causes a rapid decline in SE, but the curve levels off at a fairly small sample size (fig. 4.5A). The initial steep decline levels off at four quadrats, with a proportional SE of about 0.10. The gradual decline afterward reaches a

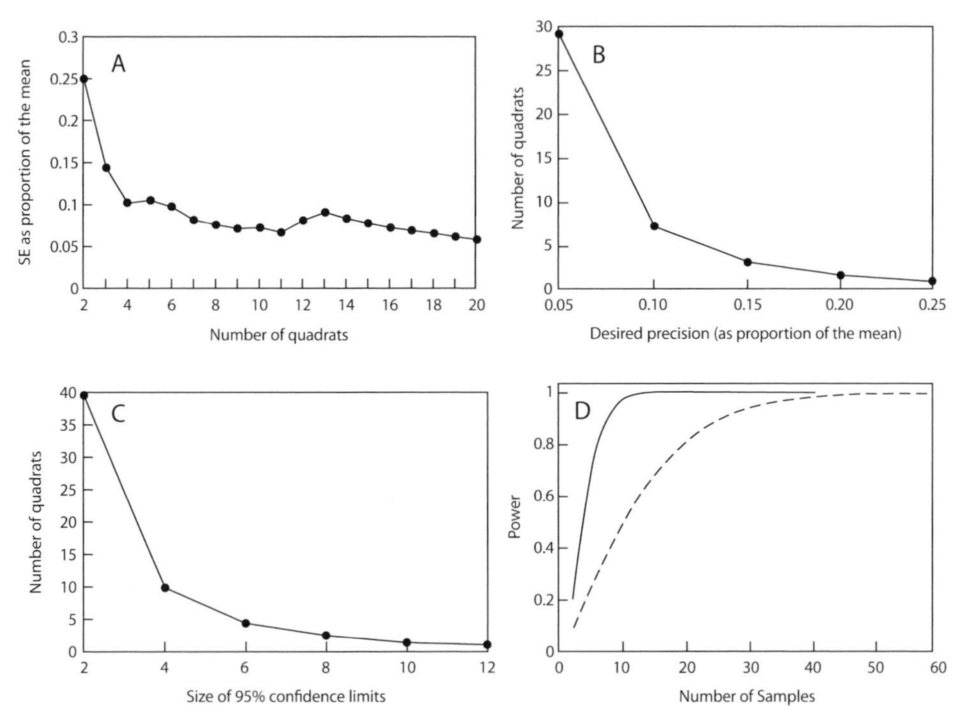

Figure 4.5. Comparison of four methods for determining appropriate sample size. Examples use hypothetical data given in table 4.2. (A) Graphical display of standard error versus sample number. (B) Number of quadrats as a function of desired precision, calculated using equation 4.1. (C) Number of quadrats as a function of the size of 95% confidence limits, calculated using equation 4.2. (D) Statistical power as a function of sample size. The solid line represents an effect size of 50% of the mean, while the dashed line represents an effect size of 25% of the mean. Power curves were calculated with $\sigma = 6.28$ and $\alpha = 0.05$. Power analyses conducted using the two-sample *t*-test from Russ Lenth's Power and Sample-Size Page; http:// www.stat.uiowa.edu/~rlenth/Power/index.html.

proportional SE of about 0.07 at 9 or 10 quadrats, with little improvement in precision thereafter. Using the values from table 4.2 and $p = 0.15$, equation 4.1 gives $[6.28/(0.15*23.3)]^2 = 3.2$ quadrats. Note that a lower desired precision, say 5% of the mean, would require the much larger sample size of 29 quadrats (fig. 4.5B). This drops to seven quadrats for 10% of the mean. Using the values from table 4.2 and an allowable error of 5 (that is, with the mean of 23.3, the 95% confidence limits are ±5), equation 4.2 gives $4 * (6.28)^2/5^2 = 6.3$ quadrats. Again, the width of the confidence limits has a large effect on the sample size required (fig. 4.5C). If the 95% confidence limits are to be ±2, then 39 quadrats will be required. This drops rapidly to nine quadrats with ±4.

Although the three approaches use different means of estimating the appropriate sample size, in this example the answers are reasonably consistent. The graphical approach indicates that 4 to 10 quadrats would be appropriate, with only a modest improvement in precision by going to 10 quadrats. Equation 4.1 indicates that three to seven quadrats would suffice, using a moderate range of the desired precision. Equation 4.2 indicates that four to nine quadrats will satisfy the criteria being employed, again using a moderate range for the 95% confidence interval.

The number of samples needed in a study also can be determined using the relationships of statistical power. Recall that there are four interrelated parameters associated with statistical power—power ($1-\beta$, the probability that a test will lead to the correct rejection of the null hypothesis), significance level (α), effect size, and variance—and that variance is a function of the variability of a sample and sample size. We can take advantage of the interrelationships of these parameters to calculate the sample size needed for a particular study. For example, if we decide we want to detect a 50% change in abundance at $\alpha = 0.05$ and 80% power, preliminary data on sample variance can be used to determine the sample size needed. If variability within the sampled population is low, fewer samples will be needed to achieve these parameter values than if variability is high. There are now many readily available computer programs that can calculate power (for a review see Thomas and Krebs 1997).

Using the power formulas with the hypothetical data given in table 4.2, we must first decide on the statistical test that will be used. For the sake of example, say we want to use a *t*-test to see if there is a significant difference in the mean abundance of a species at two sites. Assume that the variances of the two populations are equal (both 6.28) and set $\alpha=0.05$. To calculate the appropriate sample size, we must decide on the effect size (i.e., the size of the difference) we wish to detect and the power of the test. Figure 4.5D shows how power varies with different sample sizes for two different effect sizes. For example, a sample size of six is needed to detect a difference of 11.6, or half the mean of the population shown in table 4.2, with a power of 82%. A sample size of four would have only 59% power, while a sample size of nine would have 96% power. A difference of 11.6 is a rather large difference, and we might instead want to detect a difference of 5.8, or 25% of the mean—to detect a difference of this size, a sample size of four would have only 20% power. Achieving 80% power would require a sample size of 19. This is somewhat larger than the sample size estimated using the other methods.

A completely different approach to determining sample size is often taken when the parameter of concern is species richness rather than abundance. A species accumulation curve is plotted, where the cumulative number of species included in the samples is plotted on the *y*-axis and the number of samples taken is plotted on the *x*-axis. Species typically continue to accumulate, but at a decreasing rate; the appropriate number of samples is chosen somewhat subjectively as either the point at which the species accumulation curve more or less levels out or the point at which a large fraction of the species (e.g., 90%) is included.

Although the previous discussion has focused on the sample size needed to get appropriate estimates of species abundance and species richness, a similar approach can be used to determine how many samples are needed to characterize communities (Streever and Bloom 1993). Because a community can consist of many species, each with different abundances, any method for determining sample size must keep track of many variables simultaneously. One approach is to use a measure of similarity, such as Morisita's index of similarity (Morisita 1959), which incorporates information about species identities, species abundances, and species richness. A sampling effort curve can be constructed by plotting the similarity of two sample sets from the same community against progressively larger sampling efforts. Progressively greater sampling efforts yield more accurate representations of the community, so similarity increases. When the sampling effort is sufficient to represent the community, the curve levels off (typically at a value of 1).

IMPACT ASSESSMENT

A common and important purpose of environmental sampling is to assess whether a particular event or disturbance has had a significant environmental impact. There is a large literature devoted to this topic. Many of the issues discussed in this chapter need to be considered when assessing impacts. There is also the issue of the location of sampling efforts, such as which locations along a coast to sample. This question of sample locations concerns a larger scale (see chapter 2) than the issues discussed in this chapter: where to place sampling units within one location. A strong design for detecting ecological impacts is the before–after–control–impact (BACI) design and its variations. For example, one such design attempts to avoid the confounding influence of natural spatial and temporal variability by sampling a pair of impact and control sites a number of times before an impact occurs and a number of times after the impact occurs (Stewart-Oaten et al. 1986). In the absence of an impact, the difference between the control and the impact sites (called *delta;* Δ) would be expected to be the same; a significant change in delta indicates an impact.

Among the many difficulties associated with the assessment of real impacts, one of the most problematic is the fact that "before" data generally are not available or cannot be collected. Some impacts started before anyone thought to study the impacted habitats. Accidental impacts such as oil spills occur at unpredictable times and places. Although regional long-term monitoring programs may be able to provide the data necessary for a BACI analysis, appropriate data often will not exist and other approaches must be used. Wiens and Parker (1995) discuss a variety of approaches that may be possible for accidental spills, including control–impact and gradient designs (see also brief descriptions in chapter 1). Detailed discussion of these issues, which are beyond the scope of this chapter, are given by Green (1979), Stewart-Oaten et al. (1986), Underwood (1991, 1992, 1993, 1994), Schmitt and Osenberg (1996), and Kingsford and Battershill (1998).

SUMMARY

There is no single optimal sampling design for all studies. Instead, the design must be matched to the goals and constraints of each individual study. Modern ecological research in rocky intertidal communities often involves controlled experiments (Underwood 1997). Nonetheless, sampling to detect patterns of abundance or distribution, either on its own or in conjunction with experiments, remains an important part of ecological research. Sometimes, sampling is targeted on a specific component of the intertidal community, perhaps even a single species. In these cases, knowledge of the biology and distribution of that component can be used to design a specific study that usually will be strongly focused on the targeted component. Although most of the issues discussed in this chapter (such as determining the location of sampling units and the number of samples needed) still apply, the focus of such targeted studies will be narrower and the results obtained are likely to be more clear-cut compared with the outcome of a more generalized monitoring study.

On the other hand, when the study goal is to understand spatial or temporal patterns in the whole community, the decisions may not be so clear-cut. Some monitoring studies, for example, are designed to gain efficiency by identifying only target or indicator species for sampling. However, by necessity most general ecological monitoring studies must take a very broad view of their systems. This is because there generally is no way of knowing beforehand which components of the system may change or where the change might occur. The decisions to be made in designing an effective monitoring program that tracks targeted species become more difficult, too, because more trade-offs are likely to be involved. For example, many problems arise because each species has its own unique characteristics (including abundance and spatial patterns). Thus, the optimal scheme for locating sampling units is likely to differ (Miller and Ambrose 2000). And even if one scheme such as the stratified random approach can be decided on, the most appropriate way of stratifying the habitat is likely to be different for different species. Further, although a careful analysis of the appropriate sample size can be done for a species, that number will certainly differ from the number needed for another species.

Although species- and site-specific differences make it impossible to prescribe a universal sampling design, the general guidelines given in this chapter will help ensure that a sampling design is statistically rigorous and efficient. Perhaps the most important guideline is that sampling units are located in a way that ensures independence; this usually means randomizing locations. A second important guideline is to avoid placing sampling units too close together to avoid pseudoreplication. Although the most egregious pseudoreplication errors, such as placing all replicates of one treatment on one rocky intertidal bench and all replicates of another treatment on a different bench, may be obvious, care must also be taken to avoid high spatial autocorrelation from placing sampling units too close together. With the typical clumped distribution of rocky intertidal species, sampling units distributed closely in space are likely to fall in the same habitat patch. For this reason, transects, which spread sampling points along a line, generally give a more accurate estimate of species abundances at a site, particularly when they cross environmental gradients (such as tidal elevation). Finally, a conscious effort should be made at the sampling design stage to minimize variance. This often means using some kind of stratification scheme in placing sampling units in the field. In the rocky intertidal, common appropriate strata include different tidal elevations (and co-occurring species assemblages) and different habitat types (e.g., horizontal benches, crevices, and tidepools).

LITERATURE CITED

Ambrose, R.F., J.M Engle, P.T. Raimondi, J. Altstatt, and M. Wilson. 1995. Rocky intertidal and subtidal resources: Santa Barbara County mainland. Report to the Minerals Management Service, Pacific OCS Region. OCS Study MMS 95-0067.

Andrew, N.L., and B.C. Mapstone. 1987. Sampling and the description of spatial pattern in marine ecology. *Oceanogr. Mar. Biol. Annu. Rev.* 25:39–90.

Cochran, W.G. 1977. *Sampling techniques*. New York: John Wiley & Sons.

Cohen, J. 1988. *Statistical power analysis for the behavioral sciences*. 2nd ed. Hillsdale, NJ: Lawrence Erlbaum Associates.

Engle, J.M., K.D. Lafferty, J.E. Dugan, D.L. Martin, N. Mode, R.F. Ambrose, and P.T. Raimondi. 1995. Second year study plan for inventory of coastal ecological resources of the northern Channel Islands and Ventura/Los Angeles Counties. Report to the California Coastal Commission, San Francisco.

Fairweather, P.G. 1991. Statistical power and design requirements for environmental monitoring. *Aust. J. Mar. Freshwater Res.* 42:555–67.

Glass, G.V., P.D. Peckham, and J.R. Sanders. 1972. Consequences of failure to meet assumptions underlying the fixed effects analyses of variance and covariance. *Rev. Educ. Res.* 42:237–88.

Gonor, J.J., and P.F. Kemp. 1978. Procedures for quantitative ecological assessments in intertidal environments. U.S. Environmental Protection Agency Report EPA-600/3-78–087.

Green, R.H. 1979. *Sampling design and statistical methods for environmental biologists*. New York: John Wiley & Sons.

Greig-Smith, P. 1983. *Quantitative plant ecology*. 3rd ed. Berkeley: Univ. of California Press.

Hurlbert, S.T. 1984. Pseudoreplication and the design of ecological field experiments. *Ecol. Monogr.* 54:187–211.

Kingsford, M., and C. Battershill. 1998. *Studying temperate marine environments: a handbook for ecologists*. Christchurch, New Zealand: Canterbury University Press.

Mace, A.E. 1964. *Sample-size determination*. New York: Reinhold.

Mapstone, B.D. 1995. Scalable decision rules for environmental impact studies: effect size, Type I, and Type II errors. *Ecol. Appl.* 5:401–10.

Miller, A.W., and R.F. Ambrose. 2000. Optimum sampling of patchy distributions: comparison of different sampling designs in rocky intertidal habitats. *Mar. Ecol. Progr. Ser.* 196:1–14.

Morisita, M. 1959. Measuring of interspecific association and similarity between communities. *Mem. Faculty Sci. Kyushu Univ. Ser. E. Biol.* 3:65–80.

Osenberg, C.W., R.J. Schmitt, S.J. Holbrook, K.E. Abu-Saba, and R. Flegal. 1994. Detection of environmental impacts: natural variability, effect size, and power analysis. *Ecol. Appl.* 4:16–30.

Palmer, M.W., and P.S. White. 1994. Scale dependence and the species-area relationship. *Am. Nat.* 144:717–40.

Palmer, M.W., and E. van der Maarel. 1995. Variance in species richness, species association, and niche limitation. *Oikos.* 73:203–13.

Peterman, R.M. 1990. The importance of reporting statistical power: the forest decline and acidic deposition example. *Ecology.* 71:2024–27.

Richards, D.V., and G.E. Davis. 1988. *Rocky intertidal communities monitoring handbook*. Channel Islands National Park, CA: U.S. Department of the Interior.

Schmitt, R.J., and C.W. Osenberg. 1996. *Detecting ecological impacts: concepts and application in coastal habitats*. London: Academic Press.

Schroeter, S.C., J.D. Dixon, J. Kastendiek, R.O. Smith, and J.R. Bence. 1993. Detecting the ecological effects of environmental impacts: a case study of kelp forest invertebrates. *Ecol. Appl.* 3:331–50.

Snedecor, G.W. and W.G. Cochran. 1980. *Statistical Methods. 7th edition*. Ames: Iowa State University Press.

Stewart-Oaten, A. 1955. Rules and judgments in statistics: three examples. *Ecology* 76:2001–9.

Stewart-Oaten, A., W.M. Murdoch, and K.R. Parker. 1986. Environmental impact assessment: 'Pseudoreplication' in time? Ecology 67:929–40.

Streever, W.J., and S.A. Bloom. 1993. The self-similarity curve: a new method of determining the sampling effort required to characterize communities. *J. Freshwater Biol.* 8:401–3.

Thomas, L., and C.J. Krebs. 1997. A review of statistical analysis software. *Bull. Ecol. Soc. Am.* 78:126–39.

Underwood, A.J. 1991. Beyond BACI: experimental designs for detecting human environmental impacts on temporal variations in natural populations. *Aust. J. Mar. Freshwater Res.* 42:569–87.

———. 1992. Beyond BACI: the detection of environmental impacts on populations in the real, but variable world. *J. Exp. Mar. Biol. Ecol.* 161:145–78.

———. 1993. The mechanics of spatially replicated sampling programmes to detect environmental impacts in a variable world. *Aust. J. Ecol.* 18:99–116.

———. 1994. On beyond BACI: sampling designs that might reliably detect environmental disturbances. *Ecol. Appl.* 4:3–15.

———. 1997. *Experiments in ecology: their logical design and interpretation using analysis of variance.* Cambridge: Cambridge University Press.

Wiens, J.A., and K.R. Parker. 1995. Analyzing the effects of accidental environmental impacts: approaches and assumptions. *Ecol. Appl.* 5:1069–83.

Quadrat sampling at an intertidal study site near Fisherman's Cove, Santa Catalina Island, California.

CHAPTER 5

Transects, Quadrats, and Other Sampling Units

One of the most fundamental decisions that must be made concerning a sampling program is the choice of sampling units, which is the topic of this chapter. A wide variety of sampling units can be used for intertidal sampling. The most common units include line transects and plots or quadrats (fig. 5.1). In addition, plotless designs are sometimes used.

The choice of sampling unit depends on the goals of the sampling program, especially the species to be sampled. Chapter 2 addresses how to decide where to sample on the scale of study sites. Chapter 4 considers issues of sampling design, including how the sampling units should be placed at a site. This chapter discusses how to choose the type of sampling units in order to get accurate estimates of species abundances at a study site.

Sampling can be conducted using either quadrat or plot methods or plotless methods. Plots, which are the most commonly used sampling units, are discussed first. Plotless designs, which utilize some scheme to determine which elements in the environment are sampled, are discussed next. Plotless methods have rarely been used in intertidal habitats, but they might actually be the best choice for some problematic species, such as those with large individuals occurring at low densities. Finally, we consider sampling strategies for two intertidal habitats that provide special challenges, tidepools and boulder fields.

QUADRAT- OR PLOT-BASED METHODS

Quadrats and line transects are widely used sampling units. Quadrats are suitable for sampling populations with a variety of spatial patterns and densities, assuming an appropriate quadrat size (Engeman et al. 1994). As noted below, sampling plots can be labor intensive when observations are

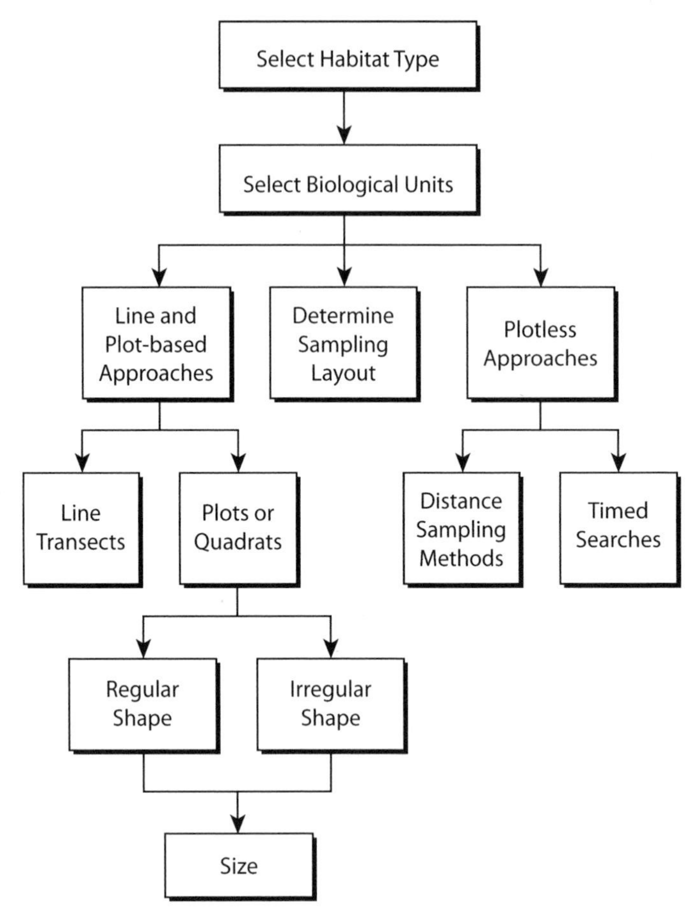

Figure 5.1. Decision tree for choosing type and size of sampling units.

sparse, unevenly distributed, or otherwise difficult to obtain (Engeman et al. 1994), and in these instances a plotless method might be more appropriate. Otherwise, plot methods will be the preferred choice for sampling intertidal organisms.

Two plot-based methods are considered here, line transects and quadrats. Line transects, which are essentially one-dimensional plots (although some transect types are considered to be plotless methods by Engeman et al. [1994]), are described first, followed by two-dimensional plots that cover an *area* of the substratum. Finally, the effectiveness of these two methods and the dimensions of sampling units are briefly compared.

Line Transects

Line transects are used primarily to estimate the cover of attached or sedentary organisms and have a long history of use in vegetation sampling (e.g., Greig-Smith 1983). There are two main approaches to line transect sampling: line intercept and point contacts. Transect lines also are often used to define bands, but these are two-dimensional sampling units and are considered below when sampling with plots and quadrats is discussed.

In the line-intercept method, the distance along the transect line that overlies each sampled biological unit or category is recorded. For example, a 5-m line-intercept transect in a surfgrass community might record sections of surfgrass interspersed with sections of rock or sand (table 5.1). Cover is determined by adding all of the segments intercepted by each category and dividing by the total line length (500 cm). In the example listed in table 5.1, surfgrass cover is 92%, rock 3%, and sand 5% along the 5-m transect line.

Line-intercept transects are relatively quick to sample (at least in some situations) and can be somewhat less subject to sampler bias (e.g., judgments about whether a species that is very close to a contact point is actually contacted) than point-contact methods. However, intercept sampling assumes a species has either 100% or 0% cover over the minimum defined intercept unit. Thus, the use of intercepts is not well suited to situations where species boundaries are not well defined or where species are mixed together. In the case of interspersed species, accurately recording large numbers of very small intercepts for each species may be

TABLE 5.1. Example of Data Collected in a Surfgrass Community Using a Line Intercept Transect Method

Species-Substration	Intercept Distance (cm)
Surfgrass	45
Rock	15
Surfgrass	150
Sand	10
Surfgrass	60
Sand	15
Surfgrass	205
Total Intercept Distance	
Surfgrass	460
Rock	15
Sand	25

NOTE: In this example, the transect is 5 m long.

Figure 5.2. Point-contact line transect. In these surfgrass transects, contacts are determined by visually recording the species directly under the contact point.

impossible. Thus, a less accurate shortcut must be taken such as deciding that the minimum unit for recording an intercept is 1 cm. Similarly, small gaps in canopy cover, which might occur when sampling sparsely distributed leaves of surfgrass (*Phyllospadix* spp.), are typically ignored, resulting in an overestimate of canopy cover.

The point-contact transect is a linear version of the point-contact methods applied to two-dimensional plots that are described in more detail in chapter 6. Point-contact transects are used more commonly in marine than in terrestrial habitats. In the standard point-contact transect, a predetermined distance along the transect serves as a point and the biological unit or category directly under this distance is recorded. The contacted item can be determined by dropping a pin or rod or by visually judging what lies directly under the contact position along the transect (fig. 5.2). The latter method is much faster but requires careful decisions about the actual position of the contact point to avoid bias such as recording a contact for a species that is close to, but not touched by, the contact point. The distance between contact points depends on the resolution needed and the length of the transect; common distances in intertidal sampling are every 10 cm and every 1 m. Random numbers also can be

TABLE 5.2. Example of Data Collected in a Surfgrass Community Using a Point-Contact Transect Method

Species	No. of Contacts
Surfgrass	46
Rock	2
Sand	2

NOTE: In this example, contacts were recorded every 10 cm along a 5-m transect.

used to determine contact point locations. Choice of sampling interval depends in part on the length of the transect, with long transects (100 m or more) having larger distances between contact points. Cover is determined by dividing the number of contacts for a particular species or sampled category by the total number of contact points recorded along the transect. Following the surfgrass example given in table 5.1, a point-contact transect sampled every 10 cm might record surfgrass under 46 points, rock under 2 points, and sand under 2 points, for an estimated cover of 92% for surfgrass, 4% for rock, and 4% for sand (table 5.2).

One advantage of transects is that they spread sampling across a large area, which can help minimize problems with spatial autocorrelation (see chapter 4 and below; Miller and Ambrose 2000). However, the traditional transect only spreads sampling along one dimension and samples a very narrow area (a line) along the second dimension. A modification of the traditional point-contact transect can be used to remedy this deficiency by spreading sampling points away from the transect line in the second dimension. In this procedure, each contact point is located by a random distance along a transect line *and* a random distance away from the transect line (e.g., at right angles to the transect, with points alternating on one side or the other of the transect line). Because this approach has elements of a point-contact transect and a point-contact quadrat, this can be referred to as a point-contact hybrid; it is analogous to a band transect sampled with point contacts. Dethier has been using this hybrid approach in rocky intertidal monitoring programs, using one transect positioned at each of four vertical elevations at each study site.

Line-intercept and point-contact transects provide very similar information about rocky intertidal habitats. Both yield estimates of percentage cover of sessile (or sedentary) organisms, and both obtain data that can be used to calculate species diversity indexes. If the *order* of collected data is maintained, both approaches can provide information describing the spatial structure of the habitat, such as whether rock coverage of 20% occurs in one large block or is dispersed in small areas or whether a species occurs at one end of the transect. In practice, spatial information is more often obtained when performing line-intercept sampling because of the logistics of data recording—maintaining the species identity and order of 50–100 point contacts is very difficult on field data sheets. However, bar code readers (battery-powered, handheld field units similar to those used in markets) can be programmed to maintain the order of point-contact data, and have the additional advantage of loading data directly into a computer database instead of manually (Ambrose and Miller, pers. observ.). Some handheld computers or personal data assistants (i.e., PDAs such as Handspring Visor and some Palm Pilots) accept a bar code reader in their expansion slots, allowing data to be scanned directly into databases that are downloaded directly to personal computers.

Plots or Quadrats

Unlike line transects, plots and quadrats have two dimensions and cover an *area* of substratum. Plots and quadrats are used to estimate cover, density, or biomass of both attached and mobile organisms (see chapter 6). The difference between quadrats and plots is somewhat semantic. Quadrats are regularly shaped and relatively small, usually of a size that can be contained within a portable quadrat frame. Plots generally are irregular in shape, although some are not, and are generally larger than quadrats.

Quadrats can be rectangular, square, or circular, and can be a wide range of sizes (see below). Relatively small ($\leq$1-m^2) quadrats are widely used in intertidal sampling programs, probably because they provide a good combination of convenience and the appropriate scale for the organisms being studied. Smaller quadrats are easily carried from sampling location to sampling location and can be easily searched or photographed for complete and thorough sampling (fig. 5.3). Quadrats and plots can be positioned randomly in an area or targeted for particular conditions; the advantages and disadvantages of these approaches are discussed in chapter 4. Regardless of how they are located, quadrats and plots can be permanently fixed at a particular place, so the same location is sampled repeatedly through time, or they can be placed in a different location during each sampling period; these approaches also are discussed in chapter 4. Quadrats are used to delimit the area to be sampled, but the particular method of sampling can vary from nondestructive counts or point contacts (chapter 6) to destructive harvesting (chapter 7).

Band transects are considered a special type of quadrat, rectangular in shape but much longer in one dimension than the other. Band transects are particularly useful for sampling large but relatively uncommon species that are distributed over large areas, such as sea stars and kelps.

Figure 5.3. Quadrat used as photoplot for sampling cover of algae and sessile animals.

Band transects can be established independent of other sampling units (fig. 5.4), or they can be sampled in conjunction with transects laid out for line-intercept or point-contact sampling, or even along transects laid out to locate other sampling units in the area. A wand (e.g., a 0.5-m or 1-m length of PVC pipe) can be used in conjunction with a transect line to determine the limits of the band. With one end of the wand held against a transect tape, the other end sets the outer limit of the band while the sampler moves along the tape. Moreover, it will be easy to decide whether most of the target organisms are clearly in or out of the band, so the wand need only be placed for organisms close to the outer band limits. The wand can be used on only one side or on both sides of the tape, but if frequent "checks" are needed, it is best to use the wand on one side only.

Another approach for estimating the densities of relatively uncommon species is to use plots. When the sampling area reaches several square meters, it is usually referred to as a plot instead of a quadrat. Plots tend to be fixed at a particular location and sampled repeatedly through time. However, there are many variations. For example, a plot might be fixed, but subsamples inside the plot are chosen at random during each sampling period (Kinnetics Laboratories, Inc. 1992).

Figure 5.4. Band transect used for sampling kelp densities. The transect is searched at a prescribed width (e.g., 0.5 m) along each side of the line using a 1.0 m wand.

Plots can be shaped to follow natural landforms and boundaries (fig. 5.5). For example, a plot for censusing sea stars can contour around a headland and encompass both flat and steep slopes within the appropriate tidal heights. Plots also are useful for sampling species that occur in particular microhabitats, such as abalone, which often occupy deep cracks and crevices. In many situations, it is difficult to achieve such a close match between appropriate habitat and the sampling unit when using quadrats or band transects. A disadvantage of irregularly shaped plots is that their irregular boundaries make it difficult to determine plot areas, a requirement if multiple plots are used to obtain density data and comparisons are to be attempted among sites. Also, it may be difficult to reconstruct plot boundaries during each sampling period, especially if plots are large and irregular in shape, and boundary markers are obscured by algae or sand. In these cases, good site maps, with distances and bearings from a variety of marked reference points, are essential, but still it can take substantial

Figure 5.5. Irregular plot used for long-term monitoring of abalone densities.

time and effort to determine plot boundaries. It is possible to combine both approaches in a sampling program, choosing regular plots or band transects when these fit the local conditions and using irregular plots when they are more appropriate for the site topography (Ambrose et al. 1995).

Little work has been done to assess the effect of different shapes of sampling units. Some "quadrats" are circular, which eliminates the extra edge effect in the corners of rectangular quadrats. In addition, a circular sampling area can sometimes be quicker and more convenient to sample compared with rectangular quadrats. For example, Ambrose et al. (1995) sampled owl limpets with a circular sampling unit whose boundaries were set by sweeping a 1-m-long line around a single, fixed stainless-steel bolt. For these sampling units, only one bolt is needed to mark each quadrat and a simple line (rather than a quadrat frame) can determine the area to be sampled. Most commonly, however, quadrats are square or rectangular. Littler and Littler (1985) report that rectangular quadrats have an advantage over square or circular plots of equal area because they may incorporate a greater diversity of populations. This advantage will be greatest when the long axis of the quadrat is oriented parallel to the environmental gradient. In rocky intertidal habitats, this gradient often corresponds with vertical tidal height so rectangular quadrats would be oriented with their long axes perpendicular to the shoreline. Often, the choice of shape may be due more to sampling logistics than sampling theory. For example, many researchers use rectangular photoquadrats with dimensions matching the framing of a 35-mm slide (e.g., 50 × 75 cm or 30 × 50 cm).

Comparing Transects and Quadrats

Quadrats are usually more appropriate for mobile species, but line transects and quadrats are both commonly used to sample seaweeds and sessile invertebrates in rocky intertidal sampling programs. The choice of sampling unit depends on the questions being asked, species to be studied, temporal dimensions of the study, and so forth. For sampling to determine species abundances, both line transects and plots or quadrats can provide accurate estimates if employed using well-designed and appropriate sampling strategies.

As mentioned in chapter 4, Miller and Ambrose (2000) used computer-simulated sampling of actual rocky intertidal data to compare point-contact transects and point-contact quadrats using different sampling designs. With one exception, transects placed perpendicular to the elevational contours and the shoreline ("vertical transects") provided more accurate estimates of overall abundance than the best quadrat efforts. This was due to the aggregated (clumped), spatial patterns displayed by species in the intertidal zone and the fact that there is an environmental gradient due to tidal ebb and flow and air exposure. In this situation, sampled points that are close to each other are more likely to yield similar values than points that are farther away. Point-contact quadrats are groupings of points, so will capture more spatial autocorrelation than point-contact transects, which spread the same number of points over a broader area.

For a relatively quick survey of the abundances of macrophytes and sessile invertebrates at a site, vertical transects running from the upper to the lower intertidal zone are probably the most efficient sampling units. Vertical transects are relatively easy to set up, especially compared with establishing strata and randomizing quadrat locations, and quick to sample. If the order of contact points is maintained, a rough map of species distributions can be produced when performing vertical transect surveys.

SIZE OF SAMPLING UNITS

Decisions about the size of a sampling unit revolve around the issue of scale. The appropriate size for a sampling unit is the size that matches the scale of the phenomenon of interest. If it is important to count the number of new barnacle recruits in the barnacle zone, where recruit density is extremely high, a 1-m^2 quadrat would be much too large. Not only would counting the high number of barnacles in this quadrat take too long, but the count would likely be inaccurate. Conversely, in most places a 1-m^2 quadrat would be too small to sample sea star or other large, low-density organisms. The appropriate quadrat size also depends on the sampling method being used. For example, barnacles may be sampled with a 0.05-m^2 quadrat when they are individually counted for density estimates but with a 0.375-m^2 quadrat when they are sampled with a point-contact method for cover estimates.

Unfortunately, sample unit sizes for rocky intertidal work have not been standardized. Gonor and Kemp (1978) compiled information on quadrat size from more than 20 rocky intertidal studies; a subset of these results, along with other more recent studies, is shown in table 5.3. Target species include coralline algae, barnacles, limpets, mussels, and snails, as

TABLE 5.3. Typical Sample Unit Sizes for Common Intertidal Species

Size (m^2)	Species	Reference
0.0225	*Littorina scutulata*	Chow 1976
0.04	*Macclintockia scabra* (as *Acmaea scabra*)	Sutherland 1970
0.06	*Laurencia papillosa*	Birkeland et al. 1976
0.0625	*Endocladia/Mastocarpus*	Kinnetics Laboratories, Inc. 1992
	Mussels	
	Miscellaneous	
0.07	Coralline algae	Littler 1971
0.10	Barnacles	Connell 1961
	Mussels	Kennedy 1976
0.125	*Littorina*	Birkeland et al. 1976
	Barnacles	Riemer 1976
	Abietinaria	
0.15	Miscellaneous	Littler and Littler 1985
0.25	*Tegula funebralis*	Frank 1975
	Gastropods	Russell 1973
	Miscellaneous	Underwood 1976
0.375	Barnacles	Richards and Davis 1988
	Turf (e.g., *Endocladia*)	Ambrose et al. 1995
	Silvetia	
	Mussels	
1.0–3.1	*Lottia gigantea*	Ambrose et al. 1995
	Leptasterias	Menge 1972
10–100	Abalone	Ambrose et al. 1995
	Sea stars	Menge 1972
		Richards and Davis 1988

SOURCES: Gonor and Kemp (1978) and other sources.

well as "miscellaneous" species, which encompassed the entire intertidal assemblage. Quadrat sizes mostly varied from 0.1 to 1 m^2, although there were a number of variations, including large quadrats (e.g., 25 m^2) that were subsampled (e.g., Boalch et al. 1974; Menge 1974; Kinnetics Laboratories, Inc. 1992).

There is no simple rule for determining the appropriate size of sampling unit for different organisms. This is not surprising given the extreme variability in the sizes and abundances of rocky intertidal organisms, especially since the same species can have widely different distributions and abundances at different sites. However, there are some general guidelines, summarized by Gonor and Kemp (1978) and described briefly here.

The two main approaches to determining sample unit size are (1) to maximize the number of species included in the sample and (2) to minimize the variance of the mean for abundance data. As the size of the sample unit increases, the sample encompasses a larger fraction of the entire area (all other things being equal), and so more and more species are likely to be included. Thus, the number of species included in a sample is expected to increase as the sample unit size increases, rather rapidly at first and then more slowly. (The same effect occurs as the number of samples taken increases.) The optimal sample unit size approximates the size where the addition of new species levels off with increases in the area sampled. This approach is appropriate when the study focus includes species richness or community-level metrics.

The second approach aims to minimize the variance of the mean associated with sampling for abundance. Minimizing variance is important because sampling programs often are concerned with detecting differences in mean abundances between sites or times, and lower variance will provide greater statistical power to detect such differences. (Other strategies for reducing variance based on the location of sampling units are discussed in chapter 4.) Sample unit size affects the variance of the mean because most, if not all, intertidal organisms are patchily distributed. For patchily distributed organisms, variance will be highest when the sample unit size matches the scale on which the patchiness occurs (Greig-Smith 1983). At sample unit sizes larger than this, variance decreases until it is approximately equal to the mean (a characteristic of random distributions). At this size, the sampling unit is so much larger than the scale of patchiness that the sample appears to have adopted a random distribution. (A similar effect occurs if the sampling unit is much smaller than the scale of patchiness [see Green 1979.]) By eliminating the extra variance caused by patchiness, the sample estimate of variance will be minimized.

The issues of sample unit size and number of sampling units used in a study are closely related since both involve sampling effort. Increasing either the size or the number of sampling units can result in more species being included in a sample and lower variances. Note that the sample unit size and the number of replicates that can be taken during a sampling period are frequently inversely related, since, all other things being equal, a larger sample unit size takes more time to sample so that fewer sample units can be included. In some cases, more accurate estimates of abundance at a site may be achieved by having more, smaller sampling units distributed throughout the area.

Finally, it is worth noting that sample unit sizes often are strongly influenced by other considerations. Smaller quadrats will be more susceptible to edge effects (see chapter 6), such as occur when observers consistently include individuals that should be excluded, and vice versa (Greig-Smith 1983). There are also important logistical and economic considerations. In most cases, larger sample units will be more expensive to sample than smaller units. Cochran (1977) discusses methods for optimizing sample unit size based on both the cost of each unit and the variance when a unit of that size is used. Logistically, some large quadrats also may be too unwieldy in the field. Furthermore, many intertidal sampling programs include a wide variety of species, and each of these could have a different "optimal" sample unit size. Although some diversity of sampling approaches is to be expected (e.g., different sampling unit sizes for attached biota and sea stars), the logistics, not to mention the extra time, of setting out and sampling different-sized sampling units for *each* species sampled would make such "optimization" infeasible. Consequently, rocky intertidal sampling programs often employ a single quadrat size to sample the abundances of most species.

PLOTLESS DESIGNS

Although quadrats and plots are commonly used for estimating species densities, sampling these can be labor intensive when observations are sparse, unevenly distributed, or otherwise difficult to obtain (Engeman et al. 1994). Plotless sampling has been introduced as an easier approach for obtaining density estimates (Cottam 1947; Persson 1971). Plotless designs have been extensively developed for vegetation sampling, particularly for forest vegetation (Greig-Smith 1983), but rarely are used for sampling intertidal populations.

Distance Sampling Methods

Many plotless approaches have been developed, most of which involve making measurements *to* individuals in the population being sampled. In a thorough evaluation of plotless approaches, Engeman et al. (1994)

compared the performance of 25 different methods using simulations of different spatial patterns, sample sizes, and population densities. The following discussion adopts the terminology given by Engeman et al. (1994), which also provides formulas for the different methods. Some of the most common and best-performing plotless methods are briefly discussed below.

In the simplest *basic distance* method, measurements are made from randomly placed sample points to the closest individual in the population (Cottam et al. 1953); variations include making measurements from individuals (usually the closest individual) to their nearest neighbors, including second-nearest neighbors, and different combinations. The *Kendall–Moran* (1963) estimation methods incorporate the total area searched for the closest individual and its nearest neighbor. The *ordered distance* method (Morisita 1957; Pollard 1971) involves measuring the distance from a random sampling point to the gth-closest individual (hence the ordering). At least for a random spatial pattern, the variance of the estimate decreases as g increases, but using $g > 3$ may be impractical in the field (Pollard 1971). *Angled-order* methods include the point-centered-quarter method, one of the oldest plotless methods. In this method, the area around a random point is divided into four quarters and the distance to the closest individual in each quarter is measured (Cottam et al. 1953). In a more general sense, the area around the random sample point is divided into k equiangular sectors and the distance to the gth-closest individual in each sector is measured (Morisita 1957). Thus, for the point-centered-quarter method, $k = 4$ and $g = 1$. Morisita (1957) considered $k = 4$ and $g = 3$ to be a practical limit. The *variable-area transect* (VAT) method is a combination of distance and quadrat methods. A fixed-width (strip or band) transect is searched from a random point until the gth individual is encountered in the strip (Parker 1979).

Engeman et al. (1994) concluded that the best-performing estimators were angled-order with $g = 3$ (AO3Q), followed by angled-order with $g = 2$ (AO2Q) (both with $k = 4$). These were followed by a Kendall–Moran estimator that pools across all sample points the search areas for the closest individual, its nearest neighbor, and the second-nearest neighbor (KM2P), an ordered-distance estimator with $g = 3$ (OD3C), and the VAT. All of these methods performed well with nonrandom spatial patterns, especially aggregated patterns, which are common in intertidal habitats. The formulas for these density estimators are given in table 5.4.

Although the best two plotless methods were angled-ordered methods, Engeman et al. (1994) agreed with Pollard (1971) that their advantages are outweighed by the practical difficulties of dividing the plane around the sampling points into quadrants and then deciding into which quadrant an individual belongs as part of the process for determining and

TABLE 5.4. Formulas for the Best-Performing Plotless Density Estimators

Estimator	Formula
Angle-order	
Third-closest individual in each quadrant	$AO3Q = 44\ N/\pi\ \Sigma\ 1/R_{(1)ij}^{\ 2}$
Second-closest individual in each quadrant	$AO2Q = 28\ N/\pi\ \Sigma\ 1/R_{(2)ij}^{\ 2}$
Kendall–Moran: CI, NN, 2N search areas pooled	$KM2P = \{[\Sigma\ (p_i + n_i + m_i)] - 1\}\ \Sigma\ C_i$
Ordered distance: third-closest individual	$OD3C = (3N - 1)/\pi\ \Sigma\ (R_{(3)i})^2$
Variable area transect	$VAT = (3N - 1)/w\ \Sigma\ l_i$

NOTE: Evaluated by Engeman et al. (1994). N—the sample size (number of random sample points used to gather distance measurements); $R_{(g)ij}$—the distance from the ith sample point to the gth closest individual in the jth sector; p_i, n_i, and m_i—the number of closest individuals, nearest neighbors, and second-nearest neighbors, respectively; C_i—the total search area at the ith sample point for the closest individual, its nearest neighbor, and the second-nearest neighbor combined; $R_{(3)i}$—the distance from the ith sample point to the third-closest individual; w—the width of the strip transect; l_i—the length searched from the random point to the gth individual.

measuring the gth-closest individuals in the quadrant. The angled-order methods also require locating more individuals at each sample point than the other methods do. Engeman et al. (1994) believe that it is usually more useful to sample more spatial points with less effort per point. For these reasons, Engeman et al. (1994) consider KM2P, OD3C, and VAT to be the three most practical plotless density estimators (PDEs) (table 5.4). Each of these methods involves locating three population individuals per random sample point. Engeman et al. (1994) note that the availability of reliable software could influence the choice of method. In particular, the algorithm for calculating KM2P is complicated, whereas it is simple to calculate OD3C and VAT.

Some plotless designs are most appropriate for conspicuous organisms that occur at low densities, such as trees. "Line-of-sight" methods, which require a clear line of sight to the objects of interest (see Buckland et al. 1993), could only be applied for conspicuous species not hidden in cracks. Even with large individuals occurring at low densities, such as sea stars, plotless designs may require a considerable amount of time searching for the appropriate "nearest" organism to measure, since all surrounding cracks and crevices would need to be examined before a decision could be made about which individual was nearest. Nonetheless, Engeman et al. (1994) consider the evaluated designs to be suitable for conditions when an area must be thoroughly searched. Plotless designs are not suitable for species with spreading cover (colonial invertebrates, most algae), since individuals cannot be easily identified, and plotless designs would be inefficient for taxa with many small individuals, such as limpets and snails.

Timed Search

The timed search is a different approach to plotless sampling that is sometimes used in intertidal sampling programs. Timed searches can be used to generate a species list for a site or to provide a quantitative or semiquantitative indication of species abundances; this approach is most applicable where target organisms are quite rare, so a very large area must be searched to encounter individuals. Timed searches are especially useful when targeted organisms occur over a large area, but suitable habitat is patchily distributed; in this way, unsuitable habitat or easily censused areas can be covered very quickly, and most search effort concentrated on the most promising habitats. Two approaches are used. In the first, a general area is identified for the search. Then this entire area is searched for a set period of time, say 30 min, and the number of individuals of the target species observed during that time is recorded. The searcher needs to be systematic, but depending on the size of the general search area, the searcher may not cover the entire area. If the site is sampled repeatedly over time, the same general area is searched each time. The second approach is similar to the first, but the area searched is measured so that density can be estimated. A systematic, thorough search of the entire area is essential in this case, so searching often starts along one boundary of the search area and proceeds systematically back and forth toward the other side of the area. This approach results in a plot whose shape and area will vary from sampling period to period and from place to place, depending on the search efficiency, environmental conditions, number of individuals encountered, and other factors. Unlike a timed search without regard for the area sampled, this approach has the advantage of allowing abundance comparisons among sites because search data can be reported as numbers of individuals per unit area (i.e., density).

The outcome of a timed search depends heavily on the skill and patience of the searcher. For rocky intertidal biodiversity surveys, the searcher must have expert knowledge of algae and invertebrates. Depth and breadth of knowledge will differ, even among experienced intertidal biologists, so comparing species lists generated by different individuals will be problematic. Between-individual differences can be minimized by using standardized checklists of common species. For estimating the abundances of targeted species, searchers with a good search image or knowledge of the habits of the species also may be much more effective at finding individuals. Thus, a fairly large discrepancy among searchers would be expected for cryptic or hidden species such as octopus and perhaps abalone, but less discrepancy would be expected for conspicuous species such as most sea stars. In addition, there is a trade-off between thoroughly searching difficult habitat and spreading the search over a broad area; individual searchers who concentrate on one or the other extreme may encounter very different numbers of targeted organisms. The limitations of timed searches can be minimized by having the same individual carry out all of the sampling. Obviously, this will not be possible with large or long-lasting monitoring programs, so comparisons among sites or sampling periods must be done very cautiously. Even where a single individual can perform all of the searches, there will be a variety of factors affecting the efficiency of searching that vary from time to time and from place to place.

Timed searches are most useful for estimating abundance when the number of target organisms expected at a site is small. In fact, the greatest utility of a timed search may be when *no* individuals are found, because it provides concrete evidence that the species was, at best, very uncommon at a site at a particular time. If the species later occurs commonly at the site, there will be good evidence for an increase in abundance. Without specific data on the abundance of a species at the site, it could be argued that it was present but simply not noted earlier. Considering that many monitoring programs could continue over decades under different personnel, or that later sampling programs might rely on data collected at a site earlier, the evidence about a particular species of interest from a timed search can be valuable in spite of the difficulties of interpreting abundance in a quantitative sense.

Timed searches also can be a relatively simple method of obtaining a more complete species list for a site than, for example, quadrat sampling alone. Typical transect and quadrat sampling efforts undersample species richness by a substantial amount (Miller and Ambrose 2000); although in theory transects and quadrats would eventually include nearly all species, the effort required to achieve this ideal would be impractical. As long as some of the caveats mentioned above are considered, species lists generated by timed searches can provide a means for comparing data among different sites.

SPECIAL HABITATS

Most of the discussion in this and other chapters focuses on rocky intertidal habitats consisting of relatively planar solid benches. Obviously, there are many other types of habitat features, even on otherwise planar benches, including cracks, deep crevices, holes, hummocks, and tidepools. These features are typically excluded from studies and ignored by not placing sampling units in them (e.g., not placing quadrats in deep crevices) or easily incorporated into a standard sampling design (e.g., counting sea stars in cracks within a large fixed plot). In this section, two of the most common types of habitats in the rocky intertidal zone that require special considerations for sampling are discussed.

Tidepools

Because of their uniqueness, tidepools are not recommended as a habitat for monitoring or impact assessment studies except where very local impacts are of interest (see chapter 2). In most rocky intertidal sampling programs, sampling units that fall largely within tidepools are reallocated to other, more appropriate locations. Tidepools must be operationally defined, such as an area of persistent water >5 cm deep, to ensure unbiased application of this rule.

If tidepools are to be sampled, the normal sampling approaches must be modified. Sampling the biota of tidepools presents some special challenges because of the three-dimensional nature and highly variable size of pools. Two basic approaches are possible: (1) each pool can be treated as one irregularly shaped plot and the biota in it sampled accordingly, or (2) sampling can be nested within each pool in an analogous way to sampling a stretch of planar bedrock.

In the first approach, the pool/plot can be surveyed using visual scans or random points to quantify algae and sessile animals, and counts of mobile organisms can be made in the whole pool or in random subsamples (Dethier 1984). Special attention must be paid to parallax problems in these three-dimensional habitats; for example, it must be determined whether the pool is to be treated as a two-dimensional plot (such as if data are being gathered from a photograph taken from above the pool) or whether the sides of the pool are to be sampled to encompass their true area. Either approach is appropriate as long as it is consistently applied. Inventories tend to focus on the amount of resource per unit area of shoreline, so a two-dimensional sampling approach for tidepools might be best for these purposes. For ecological processes, however, sampling the actual pool surface area probably will result in more useful information. The approach taken for resolving this dilemma in boulder fields is discussed below. Treating each pool as a whole plot may be most appropriate when the community of each entire pool is of concern, such as when looking at impacts of souvenir collectors or major storms.

In the second approach, each pool is treated as a stratum and then subsampled, either by running transect lines with random points through the pool or, more commonly, by placing randomly located quadrats within it (e.g., Lubchenco 1982; Benedetti-Cecchi and Cinelli 1995). This subsampling approach is more appropriate when pools are viewed as just one habitat stratum (analogous to the midintertidal zone, e.g.) in which patterns and processes are being studied.

Boulder Fields

Many rocky shores consist of unconsolidated rock with varying degrees of substratum stability. Boulder fields are often excluded from quantitative sampling in rocky intertidal monitoring efforts, perhaps because of inherent difficulties in sampling them. When boulder fields have been included in some intertidal inventories, the methods used to assess the abundances of organisms generally have not accounted for the surface relief or the undersides of boulders and, therefore, provide only limited understanding of the contribution of this habitat to intertidal resources. Gonor and Kemp's (1978) instructions for quantitative sampling on rocky shores simply recommend sampling algae and epifauna of unconsolidated beaches (consisting of gravel, cobble, or boulders) by the quadrat methods described for solid rock benches. However, Gonor and Kemp also noted that "no one method is applicable" and that the increased surface area produced by the relief of these habitats may "appreciably affect percent cover estimates as well as density." Clearly, standard methods for sampling solid rocky benches are not well suited for boulder fields.

Boulder fields present some of the same types of problems as tidepools, since boulders also are highly three-dimensional and of variable size. Proper sampling of boulder habitats requires a modification of standard rocky intertidal sampling procedures, which, especially for algae and sessile invertebrates, focus on a vertical projection for a point contact or quadrat. This is appropriate for a roughly horizontal, simple surface, such as many rocky intertidal bench habitats. However, vertical projections are less suitable for vertical surfaces and are not possible for the undersides of boulders. Thus, the standard point-contact techniques, whether along a transect or in a quadrat, are not well suited for boulder fields. Quadrats can still be used, but the methods for nondestructively sampling attached organisms in quadrats (see chapter 6) must be modified, as discussed below. Sampling mobile organisms in boulder fields is not much different from sampling on rocky benches, although boulders must be turned over to find all organisms.

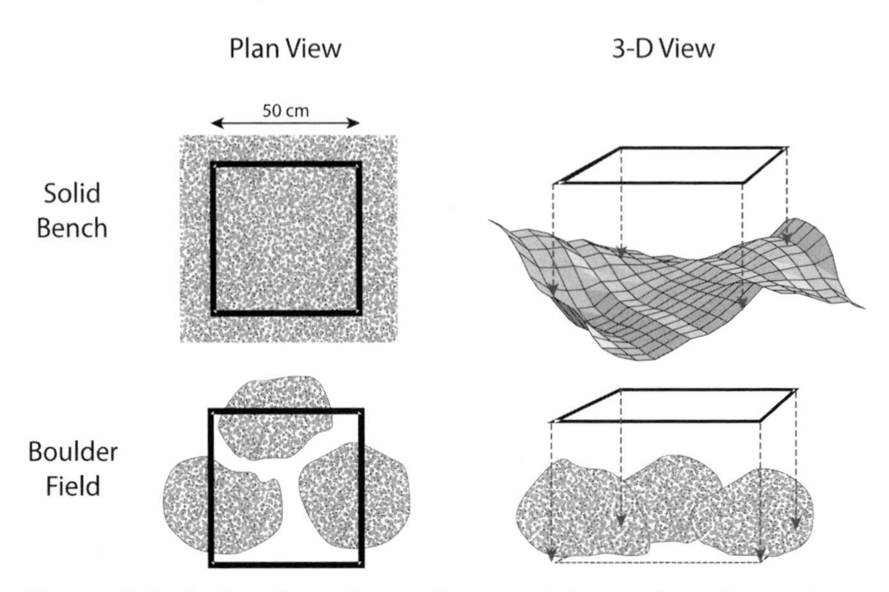

Figure 5.6. Projection of sampling quadrats onto habitat surfaces. On a rock bench, the surface area of rock under the projection is close to the nominal quadrat area, in this case $50*50 = 2{,}500\ cm^2$. In the boulder field, the surface area of rock under the projection is much greater than the nominal quadrat area because of the sides and undersides of the boulders. Moreover, cover based on the plan view (i.e., seen from above) will underestimate the abundance of species living on the sides or bottoms of the boulders.

A related issue is the way in which abundance is expressed. Common methods for determining abundances generally express results as abundance per square meter of shoreline habitat. Developed for the more or less planar solid benches of the rocky intertidal zone, these techniques are less suitable for boulder fields because they do not account for the three-dimensionality of the boulders and may also exclude organisms living in the spaces formed between or underneath boulders or in crevices. An appropriate sampling method for describing intertidal community characteristics (species richness and abundance) in boulder field habitats should account for the substratum available for attachment by fauna and flora (i.e., the *true substratum surface area*).

Pless and Ambrose (unpubl. observ.) developed an approach for sampling boulder fields in order to compare their communities with the communities on solid rock benches. The edges of a horizontally leveled sampling quadrat were vertically projected onto the substratum and the contours marked with a lumber crayon or chalk (fig. 5.6). To assess the extent of available substratum (=the true substratum surface area), the total area of rock substrata within the lumber chalk marks was estimated using small wire quadrats of different dimensions (2 × 2, 4 × 4, 5 × 5, and 10 × 10 cm) as a visual reference. Solid benches are more or less flat, so the true substratum surface area is approximately equal to the area of the sampling quadrat. The three-dimensional surface relief of boulder fields, however, can amount to a substantially larger surface area than that of a plot. The surface of boulders was measured separately by top, side, or bottom surfaces within the boundaries of a sampling quadrat. Solid benches were measured similarly, predominantly top surfaces but also sides when the surface was not entirely flat. Macroscopic organisms were sampled within the marked boundaries of the 0.25-m^2 quadrats. Estimates were determined separately for each substratum surface orientation category (top, side, bottom), which is useful for some purposes but not necessary for an inventory. Boulders were temporarily overturned to assess the abundances of organisms found on their undersides. Cover of macrophytes and sessile macroinvertebrates, such as mussels, sponges, tubeworms, tunicates, or bryozoans, was determined by visual estimates of the area covered by the individual species as a measure for biomass. Dethier et al. (1993) compared cover estimates based on visual and random-point sampling in quadrats and concluded that visual estimates gave a more accurate representation of relative coverage of sessile organisms, although both methods tended to overestimate uncommon species (see chapter 6). Visual estimates are also logistically simpler and faster than using pins or other ways of identifying contact points, particularly in boulder fields when not all surfaces are accessible from above. Considering time-effectiveness, a critical parameter in intertidal work, as well as precision and accuracy, a combination of visual estimates of cover and measurements of surface area was chosen for this study. Where a species covered less than 4 cm^2 (the smallest wire quadrat), its occurrence was simply noted and later assigned a value of 1 cm^2. Where algae occurred in several layers, each canopy was assessed separately, and thus, the total could exceed 100%.

The method described above was developed so it could be applied consistently in both boulder field and rock bench habitats. Better methods could be developed for boulder fields alone. For example, quadrats impose a flat surface of arbitrary constant size on a fundamentally three-dimensional surface of varying sizes. The standard quadrat size is particularly problematic when some boulders fit entirely within the quadrat but other boulders do not. To avoid this problem, the sampling unit could be an individual boulder. Boulders could be selected at random, perhaps stratified by size in the boulder field, and organisms on and around the boulder sampled.

SUMMARY

As with all other decisions concerning sampling, the appropriate sampling unit for a particular situation depends on the goals of the sampling program. The sampling unit determines, in part, the type of data collected. It is not possible to specify a simple prescription to be followed because each sampling program is unique. This explains at least in part the great diversity of sampling units that have been used in rocky intertidal studies. However, some general considerations are presented in this section.

For sampling that involves counts of individuals, quadrats or plots are usually used. Most commonly, a rectangular or square quadrat is appropriate, with the size dependent on the density of individuals being sampled and logistics. When the density is low, a plot or band transect is needed, or else a very large number of smaller quadrats is required to obtain an accurate estimate of abundance. Plotless methods rarely have been used in rocky intertidal studies but should be considered for larger species such as sea stars and abalone.

The placement of sampling units in a study area was considered in detail in chapter 4, but one aspect is relevant to the choice of sample unit. When program goals include a description of the abundances of a variety of species, a long, linear sampling unit (such as a line or band transect) that crosses different patch types or environmental gradients will provide the most accurate and efficient estimate of mean abundances (Clapman 1932; Bormann 1953; Miller and Ambrose 2000). The linear sampling unit spreads the sampling effort over a greater portion of the site and, thus, is likely to include samples taken in different habitat patches.

When repeated samples at a site are desired to track the performance of specific individuals or assemblages, permanent or fixed plots can be used to ensure that all samples include the same spot. Fixed transects can be used for sampling the same general area, but it is difficult to establish transects precisely enough that the exact same spots can be repetitively sampled. Fixed quadrats, on the other hand, can be relocated with great precision.

Finally, it is worth noting that the discussion of sampling units in this chapter concerns relatively conspicuous macroalgae and macroinvertebrates. Most general studies of rocky intertidal communities and impacts on these communities will have a similar focus. However, specific, targeted studies may require different sampling approaches. For example, cryptic species, especially those occurring in holes or crevices or boring in rock, may not be sampled well with the sampling units described here. In other cases, natural sampling units may be more appropriate. For example, species occurring in kelp holdfasts may be sampled using holdfasts as the sampling unit (Jones 1973), and organisms occurring in intertidal boulder fields may be best sampled using individual boulders as the sampling unit (Pless and Ambrose, unpubl. observ.).

LITERATURE CITED

Ambrose, R.F., J.M. Engle, P.T. Raimondi, M. Wilson, and J.A. Altstatt. 1995. Rocky intertidal and subtidal resources: Santa Barbara County mainland. Report to the Minerals Management Service. Pacific OCS Region. OCS Study MMS 95-0067.

Benedetti-Cecchi, L., and F. Cinelli. 1995. Habitat heterogeneity, sea urchin grazing and the distribution of algae in littoral rock pools on the west coast of Italy (western Mediterranean). *Mar. Ecol. Progr. Ser.* 126:203–12.

Birkeland, C., A.A. Reimer, and J.R. Young. 1976. Survey of marine communities in Panama and experiments with oil. U.S. Environmental Protection Agency Report EPA-600/3-76-028, Narragansett Laboratory.

Boalch, G.T., N.A. Holme, N.A. Jephson, and J.M.C. Sidwell. 1974. A re-survey of Colman's intertidal traverses at Wembury, South Devon. *J. Mar. Biol. Assoc. UK* 54:551–53.

Bormann, F.H. 1953. The statistical efficiency of sample plot size and shape in forest ecology. *Ecology* 34:499–512.

Buckland, S.T., D.R. Anderson, K.P. Burnham, and J.L. Laake. 1993. *Distance sampling: estimating abundance of biological populations*. London: Chapman and Hall.

Chow, V. 1976. The importance of size in the intertidal distribution of *Littorina scutulata*. *Veliger* 18:69–78.

Clapman, A.R. 1932. The form of the observational unit in quantitative ecology. *J. Ecol.* 20:192–97.

Cochran, W.G. 1977. *Sampling techniques*. 3rd ed. New York: John Wiley & Sons.

Connell, J.H. 1961. The influence of interspecific competition and other factors on the distribution of the barnacle *Chthamalus stellatus*. *Ecology* 42:710–23.

Cottam, G. 1947. A point method for making rapid surveys of woodlands. *Bull. Ecol. Soc. Am.* 28:60.

Cottam, G., J.T. Curtis, and B.W. Hale. 1953. Some sampling characteristics of a population of randomly dispersed individuals. *Ecology* 34:741–57.

Dethier, M.N. 1984. Disturbance and recovery in intertidal pools: maintenance of mosaic patterns. *Ecol. Monogr.* 54:99–118.

Dethier, M.N., E.S. Graham, S. Cohen, and L.M. Tear. 1993. Visual vs. random-point percent cover estimations: 'objective' is not always better. *Mar Ecol Progr. Ser.* 96:93–100.

Engeman, R.M., R.T. Sugihara, L.F. Pank, and W.E. Dusenberry. 1994. A comparison of plotless density estimators using Monte Carlo simulation. *Ecology* 75:1769–79.

Frank, P.W. 1975. Latitudinal variation in the life history features of the black turban snail *Tegula funebralis*. *Mar. Biol.* 31:181–92.

Gonor, J. J., and P. F. Kemp. 1978. Procedures for quantitative ecological assessments in intertidal environments. U.S. Environmental Protection Agency Report EPA-600/3-78-087.

Green, R. H. 1979. *Sampling design and statistical methods for environmental biologists*. New York: John Wiley & Sons.

Greig-Smith, P. 1983. *Quantitative plant ecology*. 3rd ed. Berkeley: Univ. of California Press.

Jones, D. L. 1973. Variation in the trophic structure and species composition of some invertebrate communities in polluted kelp forests in the North Sea. *Mar. Biol.* 20:351–65.

Kendall, M. G., and P. A. P. Moran. 1963. *Geometrical probability*. London: Griffin.

Kennedy, V. S. 1976. Desiccation, higher temperatures, and upper intertidal limits of three species of sea mussels in New Zealand. *Mar. Biol.* 35:127–37.

Kinnetics Laboratories, Inc. 1992. Study of the rocky intertidal communities of central and northern California. Final Report KLI-R-91-8 to the Minerals Management Service, Pacific OCS Region. OCS Study MMS 91–0089.

Littler, M. M. 1971. Standing stock measurements of crustose coralline algae (Rhodophyta) and other saxicolous organisms. *J. Exp. Mar. Biol. Ecol.* 6:91–99.

Littler, M. M., and D. S. Littler. 1985. Nondestructive sampling. In *Handbook of phycological methods. Ecological field methods: macroalgae*, ed. M. M. Littler and D. S. Littler, 161–75. Cambridge: Cambridge Univ. Press.

Lubchenco, J. 1982. Effects of grazers and algal competitors on fucoid colonization in tide pools. *J. Phycol.* 18:544–50.

Menge, B. A. 1972. Foraging strategy of a starfish in relation to actual prey availability and environmental predictability. *Ecol. Monogr.* 42:25–50.

———. 1974. Prey selection and foraging period of the intertidal predaceous snail *Acanthina punctulata*. *Oecologia* 17:293–316.

Miller, A. W., and R. F. Ambrose. 2000. Optimum sampling of patchy distributions: Comparison of different sampling designs in rocky intertidal habitats. *Mar. Ecol. Progr. Ser.* 196:1–14.

Morisita, M. 1957. A new method for the estimation of density by spacing method applicable to nonrandomly distributed populations. *Physiol. Ecol.* 7:134–44. (In Japanese; available as Forest Service translation no. 11116, USDA Forest Service, Washington, DC.)

Parker, K. R. 1979. Density estimation by variable area transect. *J. Wildlife Manage.* 43:484–92.

Persson, O. 1971. The robustness of estimating density by distance measurements. In *Statistical ecology, Vol.* 2, ed. G. P. Patil, E. C. Pielou, and W. E. Waters, 175–87. University Park: Pennsylvania State University Press.

Pollard, J. H. 1971. On distance estimators of density in randomly distributed forests. *Biometrics* 27:991–1002.

Richards, D. V., and G. E. Davis. 1988. *Rocky intertidal communities monitoring handbook*. Channel Islands National Park, CA: U.S. Department of the Interior.

Riemer, A. A. 1976. Description of a *Tetraclita stalactifera panamensis* community on a rocky shore of Panama. Mar. Biol. 35:225–38.

Russell, G. 1973. The litus line: a reassessment. *Oikos* 24:158–61.

Sutherland, J. P. 1970. Dynamics of high and low populations of the limpet *Acmaea scabra* (Gould). *Ecol. Monogr.* 40:169–88.

Underwood, A. J. 1976. Analysis of patterns of dispersion of intertidal prosobranch gastropods in relation to macroalgae and rockpools. *Oecologia* 25:145–54.

Lower intertidal kelp and coralline algal communities at Shaw's Cove, Laguna Beach, California.

CHAPTER 6

Quantifying Abundance

Density and Cover

Most rocky intertidal studies require quantification of abundance, of either individual species or other taxonomic units of the investigator's choice. Data indicating species abundances can be collected quickly using subjective scales or by determining the presence of species in defined sampling units. Robust studies of species abundance, however, require more objective approaches that extend beyond subjective estimates or simple presence or absence determinations. Three kinds of quantitative data are used to express the abundances of rocky intertidal organisms in these more powerful studies: numerical counts or density, percentage cover, and biomass. The particular method of quantifying abundance will vary with the nature of the sampled taxa and the specific goals of the study. Counts are used most often for expressing the abundances of mobile animals and larger seaweeds (e.g., kelps) where individual genets can readily be distinguished. Percentage cover is commonly used for quantifying the abundances of most seaweeds, colonial organisms, such as sponges and bryozoans, and many species of sessile invertebrates including barnacles, mussels, and tube-dwelling worms and mollusks. Alternatively, the abundances of macroalgal and macroinvertebrate populations can be quantified in terms of biomass. Biomass data might be favored or required if investigators are interested in the energetic contributions of populations to an intertidal community.

This chapter reviews various methods of quantifying the density and cover of rocky intertidal populations using nondestructive sampling procedures. Methods for determining biomass, which usually require that organisms be extracted and returned to the laboratory, are considered in chapter 7. Procedures used to collect abundance data during rapid surveys and while using plotless and line transect sampling methods are briefly discussed. Emphasis in this chapter, however, has been placed on

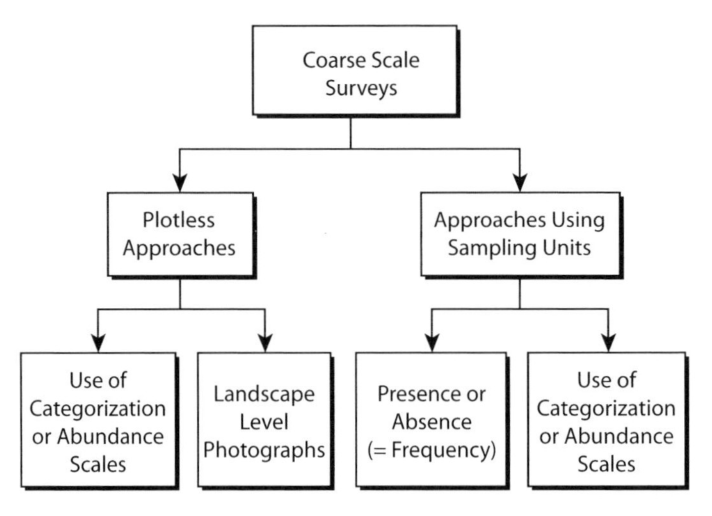

Figure 6.1. A summary of approaches for obtaining data when performing rapid surveys of rocky intertidal populations.

plot- or quadrat-based methods for determining density and cover of intertidal macroalgae and macroinvertebrates. As discussed in chapter 5, differences between plots and quadrats are somewhat semantic and the terms are used as functional synonyms in this chapter. Additional discussions of plotless and line transect sampling approaches can be found in chapter 5.

RAPID SURVEYS

The goals of some studies might not require quantitative estimates of numerical abundance, cover, or biomass and can be met by less rigorous and less labor-intensive sampling procedures (fig. 6.1). Subjective or semiquantitative approaches are most useful for broad-scale surveys (e.g., Crisp and Southward 1958; Dawson 1959, 1965; Boudouresque 1969, 1971; Baker and Crothers 1987), where they offer the advantage of providing rapid, integrated estimates of species abundances in often-patchy environments. Subjective data can be collected in such rapid surveys by making scaled abundance estimates for species integrated over large sections of habitat or by applying scalar approaches to discrete sampling units such as quadrats distributed along transect lines.

For example, in his surveys of 44 rocky intertidal sites, Dawson (1959, 1965) subjectively assigned species to one of three categories (rare to

TABLE 6.1. The Braun–Blanquet System for Providing Semi-quantitative or Subjective Assessments of Vegetation Abundance

Scale	Description of Abundance and Growth Form
+	Sparsely or very sparsely present; cover small
1	Plentiful but of small cover
2	Very numerous, or covering at least 5% of the area
3	Any number of individuals, covering 25% to 50% of the area
4	Any number of individuals covering 50% to 75% of the area
5	Covering more than 75% of the area
Soc 1	Growing singly, isolated individuals
Soc 2	Growing grouped or in tufts
Soc 3	Growing in small patches or cushions
Soc 4	Growing in small colonies, or in extensive patches, or forming a carpet
Soc 5	Growing in pure populations

SOURCE: After Kershaw (1973).

NOTE: The first six rows provide information on abundance, whereas the second five rows describe sociability (Soc) or growth form.

scant, occasional to frequent, and common to abundant) based on field observations performed over undefined areas extending out from a fixed transect line. More defined qualitative observations also can be used to assign species to arbitrary abundance categories or scales (e.g., ACFOR: abundant, common, frequent, occasional, or rare) using quadrats. One such approach was developed by Braun-Blanquet (1927), who proposed two scales for categorizing subjective observations of vegetation abundance, one for combining the number and cover of a species and the other providing a measure of sociability or growth pattern (table 6.1). Another approach, described by Hawkins and Jones (1992), relies on seven subjective abundance categories and defines the criteria for each category for each type of intertidal organism (table 6.2). This scheme is an extension of the five-category system developed by Crisp and Southward (1958) in their broad-scale study of the biogeography of rocky shore species in the English Channel.

Quick surveys also can be performed by determining only the presence or absence of a taxon in defined samples such as quadrats or intervals distributed along transect lines. These data can then be used to calculate frequency (i.e., the proportion of the total number of sample units within which a given species has been recorded), a representation of how common the taxon is in the study area. Frequency data are automatically derived from any quantitative sampling program where sampling units

TABLE 6.2. Abundance Scale Categories for Different Groups of Rocky Shore Organisms from the North Atlantic

Seaweeds	Lichens and Lithothamnia Crusts (e.g., Crustose Corallines)	Small Barnacles	Whelks, Topshells, Anemones, and Sea Urchins	Limpets
E = >90% cover	E = >80% cover	E = >5 cm^{-2}	E = >100 m^{-2}	E = >200 m^{-2}
S = 60%–90% cover	S = 50%–79% cover	S = 3–5 cm^{-2}	S = 50–90 m^{-2}	S = 100–200 m^{-2}
A = >30% cover	A = >20% cover	A = >1 cm^{-2} (rocks well covered)	A = >10 m^{-2}	A = >50 m^{-2}
C = 5%–30% cover	C = 1%–20% cover (zone well defined)	C = 0.1–1 cm^{-2} (up to 1/3 of rock covered)	C = 1–10 m^{-2}; very locally >10 m^{-2}	C = 10–50 m^{-2}
F = <5% cover (zone apparent)	F = <5% cover (zone ill defined)	F = 100–1,000 m^{-2} (individuals never >10 cm apart)	F = <1 m^{-2}; locally sometimes more	F = 1–10 m^{-2}
O = Scattered individuals (zone indistinct)	O = Small widely scattered patches	O = 1–100 m^{-2} (few within 10 cm of each other)	O = Always <1 m^{-2}	O = <1 m^{-2}
R = Few plants	R = Few patches seen	R = Few found	R = 1 or 2 found	R = Few found

Large Barnacles	*Littorina littorea*	Mussels and *Sabellaria*	Littorines and Periwinkles	Tube Worms: *Pomatoceros*	Tube Worms: Spirorbids
E = >300 per 10 × 10 cm	E = >200 m^{-2}	E = >80% cover	E = >5 cm^{-2}		
S = 100–300 per 10 × 10 cm	S = 100–200 m^{-2}	S = 50–79% cover	S = 3–5 cm^{-2}		
A = 10–100 per 10 × 10 cm	A = >50 m^{-2}	A = >20% cover	A = >1 cm^{-2} at HWN (extending down to midshore)	A = >500 m^{-2}	A = 5 cm^{-2} on > 50% of surfaces
C = 1–10 per 10 × 10 cm	C = 10–50 m^{-2}	C = Large patches	C = 0.1–1 cm^{-2} (mainly in littoral fringe)	C = 100–500 m^{-2}	C = 5 cm^{-2} on < 50% of surfaces
F = 10–100 m^{-2}	F = 1–710 m^{-2}	F = Scattered individuals/ small patches	F = <0.1 cm^{-2} (mainly in crevices)	F = 10–100 m^{-2}	F = 1–5 cm^{-2}
O = 1–9 m^{-2}	O = <1 m^{-2}	O = Scattered individuals/ no patches	O = A few individuals in deep crevices	O = 1–9 m^{-2}	O = <1 cm^{-2}
R = Few found	R = 1 or 2 found	R = Few seen	R = 1 or 2 found	R = <1 m^{-2}	R = Few found

SOURCE: After Hawkins and Jones (1992).

NOTE: Organism categories listed are for the United Kingdom and need to be modified for other regions. Scales: E, extremely abundant; S, superabundant; A, abundant; C, common; F, frequent; O, occasional; R, rare; N, not found (all cases). R assignments based on 30-min search.

are identified and density, cover, or biomass data are taken. Finally, quantitative cover data can be obtained efficiently in surveys, particularly for space-occupying macrophytes and sessile invertebrates that are easy to identify, by recording point contacts along transect lines (see chapter 5 for discussion of point-contact sampling procedures). Transect lines also can be used to perform rapid video surveys, with data being extracted in the laboratory by standard line sampling procedures. However, as discussed later in this chapter, resolution and problems resulting from three-dimensional algal canopies and invertebrate assemblages will limit the ability to collect data. Nevertheless, video transects can be used effectively in rapid surveys to quantify the cover of easy-to-identify, habitat-structuring intertidal populations.

The use of semiquantitative abundance data will be limited and its collection is recommended only when coarse-scale descriptions of species distributions and abundances are required (Creese and Kingsford 1998). This is because subjective abundance estimates can vary considerably from investigator to investigator and uncertainties can exist about the specific habitats assessed (e.g., whether the estimated abundance value for a lower intertidal kelp is based on its perceived abundance for the entire intertidal zone or just for the lower shoreline). Semiquantitative abundance data also create problems in statistical analysis and make it difficult to detect changes in populations over long periods of time (Raffaelli and Hawkins 1996).

PLOTLESS AND LINE TRANSECT METHODS

Various plotless and line transect methods can be used to obtain density or cover data in intertidal sampling programs (fig. 6.2). These include distance-based techniques and mark-recapture methods for density and the use of multiple points scattered randomly or systematically throughout the study area and points or intercept lengths along transect lines for cover. Terrestrial plant ecologists have largely developed and used plotless methods for determining species densities. A discussion of these approaches is presented in chapter 5. Although plotless techniques seldom have been employed in intertidal studies, Loya (1978) has described their use for sampling coral reef populations. Mark–recapture methods most often are used to obtain density data on highly mobile animals, such as vertebrates, where direct counting is impractical. However, mark-recapture techniques generally are not applicable for studying intertidal invertebrates, most of which are small and slow-moving, except for large, highly mobile crustaceans such as crabs and lobsters. Unfortunately, even for these organisms, mark–recapture methods are difficult to perform because of tagging (see chapter 8) and observational difficulties.

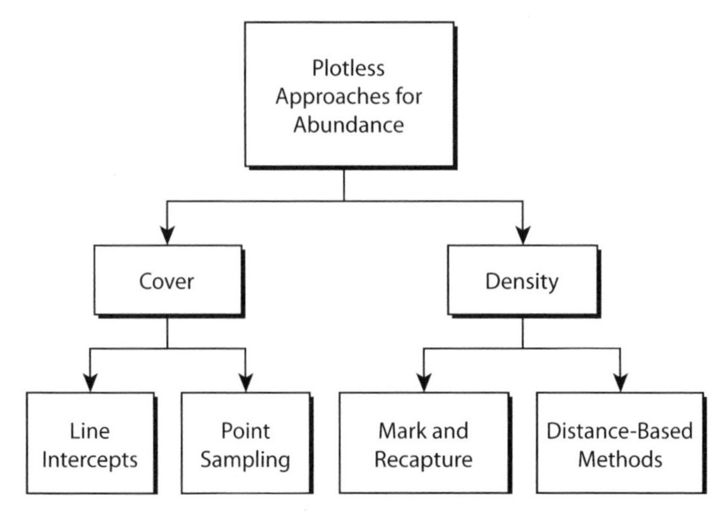

Figure 6.2. A summary of nondestructive, plotless approaches for obtaining abundance data on rocky intertidal populations.

Cover of intertidal organisms is usually measured using plots or quadrats, although cover also can be estimated without using plot-based sampling units. For example, points can be positioned randomly throughout the study area using transect tapes or other methods or can be regularly spaced along randomly located transect lines. As discussed in chapter 5, point sampling along transect lines can be an efficient and repeatable procedure for obtaining cover estimates of rocky intertidal organisms during broad-scale surveys or focused monitoring or scientific studies. Multiple transect lines can be used to collect point contacts, with each line serving as a replicate to provide the statistical advantage of multiple sampling units. Although a lengthy record exists for employing line intercept methods to determine the cover of terrestrial vegetation (Greig-Smith 1983), this form of sampling has been used much less frequently in intertidal work, particularly during recent years. Additional discussions of distance-based techniques, line intercept procedures, and point sampling can be found in the previous discussion of transects, quadrats, and other sampling units (chapter 5). A discussion of the use of mark–recapture methods for estimating population densities is given by Krebs (1989).

PLOT- OR QUADRAT-BASED METHODS

Plot- or quadrat-based methods for determining abundance are emphasized in the discussions presented here and in chapter 7. These methods

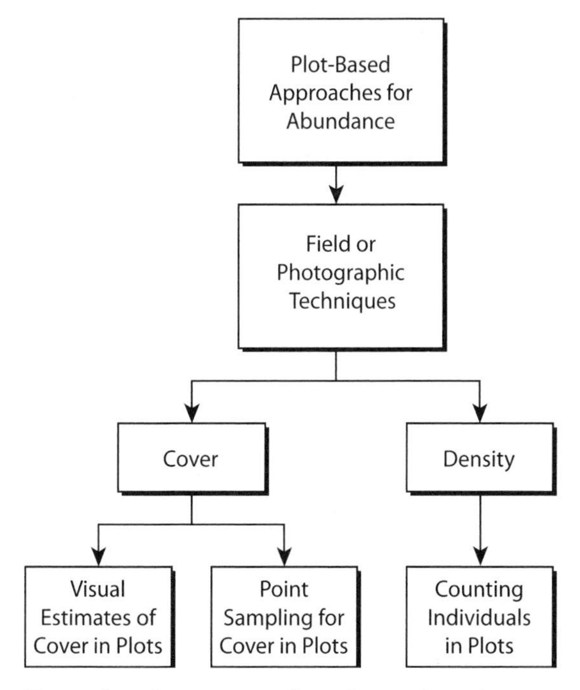

Figure 6.3. A summary of nondestructive, plot-based approaches for obtaining abundance data on rocky intertidal populations.

are employed commonly to quantify the cover or density of macroorganisms in rocky intertidal monitoring, impact, baseline, and experimental studies, particularly those that require repeated assessments (fig. 6.3). Gonor and Kemp (1978), Littler and Littler (1985), and Creese and Kingsford (1998) provide good previous accounts of nondestructive sampling methods for quantifying the abundances of intertidal populations. Other discussions of sampling topics treated in this chapter are given by Kershaw (1973), Mueller-Dombois and Ellenberg (1974), Greig-Smith (1983), Andrew and Mapstone (1987), Krebs (1989), Hayek and Buzas (1997), Hawkins and Jones (1992), Brower et al. (1998), and Kingsford and Battershill (1998).

Density

Direct counts of the number of individuals is perhaps the most intuitive means of expressing abundance and is clearly the most commonly used, nondestructive method for sampling the abundances of mobile intertidal invertebrates such as limpets, turban snails, crabs, littorines, and chitons. To facilitate comparisons with other studies where quadrat sizes or the area being sampled differ, counts of organisms should be converted to density or the number of individuals per unit area of intertidal surface (preferably per square meter). Most investigators use regularly shaped quadrats for sampling intertidal seaweeds and macroinvertebrates, although large, irregularly shaped areas may be necessary for sampling certain, patchily distributed organisms such as sea stars and abalone (see chapter 5 and Ambrose et al. 1992). Numerical counts not only provide a useful parameter for expressing abundance but also, together with size-frequency data, can form the basis for describing demographic attributes of populations. Numerical counts also are useful for calculating various indexes that describe community structure and composition (e.g., diversity indexes [Magurran 1988]) or that compare communities (e.g., similarity indexes [Pielou 1984]) based on species abundances.

Obtaining accurate counts requires the ability to discern individuals, a task that is difficult, if not impossible, for most seaweeds and many encrusting animals (e.g., sponges and colonial ascidians). Similar problems have long been recognized by terrestrial ecologists interested in sampling rhizomatous or stoloniferous herbs and mosses (Kershaw 1973; Mueller-Dombois and Ellenberg 1974; Greig-Smith 1983). Despite the conceptual simplicity of counting organisms in plots or quadrats, this task can be time-consuming and difficult to employ, particularly when organisms are small and numerous, such as acorn barnacles or smaller species of littorine snails. Accurate counts also can be difficult to obtain for abundant organisms with scattered spatial distributions because of difficulties in remembering which individuals have been tallied during the counting process. To keep track of counted individuals, organisms can be marked as they are encountered using a wax pencil, lumber crayon, or chalk. Use of a handheld mechanical counter can speed up the counting process when organisms are abundant (>50). Smaller quadrats can be processed more efficiently and accurately than larger ones when counting small, very abundant organisms. Alternatively, smaller quadrat sections can be subsampled when using quadrats with dimensions better designed for sampling larger, less numerous species. The use of more than one quadrat size should be considered in the design of population-based studies where there are substantial differences in the sizes and densities of the organisms being sampled (Kingsford and Battershill 1998).

All plots or quadrats have boundaries and difficulties arise in determining when to include or exclude organisms positioned across plot borders. This problem is exacerbated by observer parallax because the angle at which the plot is viewed in the field actually sets its boundaries. Hence, observers should attempt to reduce parallax error by viewing plot or quadrat boundaries at an angle perpendicular to the substratum. Using plots or quadrats with a lower perimeter-to-area ratio (Krebs 1989) also can reduce these edge effects. Boundary problems can bias the counting process and result in overestimates of numerical abundance, even by experienced investigators who can be tempted to include all organisms located along plot borders. To eliminate bias, consistent procedures should be established for counting organisms found on plot boundaries. These procedures or sampling rules must not result in the counting of all organisms located along plot borders because to do so would be equivalent to uniformly expanding the plot area.

Sampling rules for plot borders should be developed *a priori* to avoid bias when performing counts (fig. 6.4). For example, the proportion of an organism located inside plot or quadrat boundaries can be used to decide whether or not the organism should be counted. This rule is most easily applied for large, regularly shaped organisms, where an individual is counted if it resides at least 50% within the plot (Gonor and Kemp 1978). For sessile animals and plants, inclusion should be based only on the attachment area or holdfast; organisms attached outside plot boundaries but with body parts or fronds that lie within plot borders should not be included in plot counts. For organisms where size and shape make it difficult to determine the proportion included within the plot, individuals can be omitted from counts if they are positioned across two of a plot's four sides (Gonor and Kemp 1978). For rectangular plots, the two borders should include one large and one small side so that the sides selected bias neither inclusion nor exclusion. The specific sides of the plot to be excluded should be identified prior to counting and held consistent throughout the study to simplify sampling and avoid the temptation of making adjustments that result in the unwarranted inclusion of rare species in plot counts.

Cover

Although density data usually can be collected for mobile invertebrates, kelps, and other large algae, it is usually impossible to discern and count individuals of most seaweeds and colonial invertebrates such as sponges and bryozoans. Discriminating individual seaweeds or genets is often impossible because holdfasts of neighboring thalli can grow tightly together and even fuse, and because cryptic heterotrichous basal systems can give rise to multiple, upright fronds (fig. 6.5). Interestingly some red algal germlings can coalesce during early development, resulting in the production of multiple chimeric fronds from a single basal system (Maggs and Cheney 1990; Murray and Dixon 1992). For colonial invertebrates, quantifying the numbers of colony members or colonies might be less

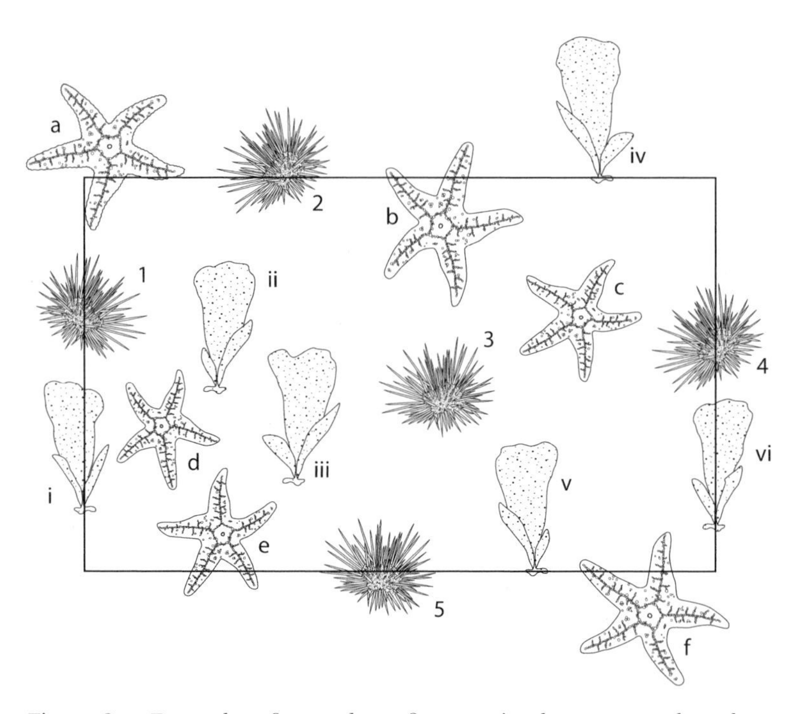

Figure 6.4. Examples of procedures for counting large seaweeds and macroinvertebrates in field plots. For seaweeds and sessile animals, only individuals attached to surfaces within plot boundaries are counted when taking density data. For mobile animals, *a priori* sampling rules should be used to determine whether an individual is to be counted. For example, for large, regularly shaped animals it may be decided *a priori* that an individual will be counted if 50% or more of its body lies inside plot boundaries. For small sessile or mobile animals and small seaweeds, organisms often are included in counts if they touch two of the four plot borders but are not included if they touch the other two. Based on the former counting strategy, sea stars a and f are not included in counts because more than 50% of their bodies lie outside plot boundaries. If the latter counting strategy is to be employed, urchins 1 and 2 and seaweeds i and iv are not included in counts because they lie across the upper and left plot borders, which have been designated for exclusion.

important than measuring the area of the occupied substratum. Sessile barnacles, mussels, and tube-dwelling worms and mollusks usually can be discriminated in the field, but these organisms often are so numerous that counts can be extremely time-consuming and of less ecological importance than the quantification of occupied space (fig. 6.6). Hence, investigators generally use percentage cover to express the abundances of most

Figure 6.5. Coralline algal turf consisting of *Lithothrix aspergillum* and *Corallina pinnatifolia* with multiple upright axes arising from an uncertain number of individual crustose bases. For many seaweeds, it is impossible to determine the actual number of discrete individuals that make up a patch or clump.

seaweeds and colonial invertebrates. Similarly, the abundances of many sessile invertebrates often are reported as percentage cover unless the study goals require density data.

Cover is expressed as the percentage (or proportion) of the surface area of the sample unit (e.g., plot) covered by invertebrate bodies or seaweed thalli. In intertidal studies, percentage cover also is commonly used to quantify the amount of sand, tar, unoccupied rocky substratum, or substratum type. Unlike the previously described procedures for counting organisms, percentage cover estimates usually are made on all material overlying the perpendicular projection of the substratum area contained within plot borders. Thus, when using plots, cover estimates should be made from a viewing angle perpendicular to the substratum to standardize the plot area to that approximating its planar surface. This procedure will make the cover of a plot occurring on a horizontal substratum equivalent to the cover of the same-sized plot placed against a vertical wall when viewed normal to the wall's surface, irrespective of the substratum

Figure 6.6. Dense aggregation of barnacles occupying a flat, rocky surface in the upper intertidal zone. It is often difficult to obtain accurate counts of small, densely aggregated barnacles and other invertebrates that occur at densities of several thousand per square meter.

topography. Nevertheless, whether plots are positioned horizontally or vertically, the actual surface areas of the substrata being sampled will almost surely vary among plots because of differences in topographical relief. Unless the goals of the study require more precise treatment of topographical heterogeneity (e.g., see Trudgill 1988; Underwood and Chapman 1989; Kingsford and Battershill 1998), however, investigators usually consider all plots of the same dimensions to sample equivalent areas. Since the planar area of a plot can be calculated from its dimensions, this area can be used to convert percentage cover to metric data when this is required.

A common difficulty in estimating the cover of intertidal organisms is layering or multiple occupancy of the space directly above the primary substratum (fig. 6.7). Layering can be due to overlapping parts of the same individual or the shared occupancy of the vertical extension of the primary substratum by more than one organism. For example, erect fronds of most seaweeds collapse during low tide, leaving the entire alga

Figure 6.7. Layering is common in intertidal habitats occupied by large, canopy-forming seaweeds such as *Eisenia arborea* in southern California.

lying horizontally across the rock surface. Although the fronds of many temperate intertidal seaweeds usually are less than 10–20 cm long, kelps and other large algae can have fronds that reach lengths greater than 1 m. Thus, the fronds of larger seaweeds form canopies that cover other organisms during low tide. Layering also occurs at smaller scales when smaller algae or invertebrates occur as epibionts on seaweeds or on the shells of organisms. Thus, cover is stratified vertically on most shores but generally at scales of only tens of centimeters, except where larger algal canopies are present.

Although layering is common in seaweed-dominated communities, sessile invertebrates also can contribute to the three-dimensional structure of intertidal communities. For example, barnacles provide substratum for other organisms and mussels can form thick, layered beds that offer attachment surfaces and living space for numerous invertebrates and seaweeds. Most researchers attempt to include all of these organisms when making cover estimates instead of recording only the uppermost, visible layer. This is because failure to include the cover of all taxa occupying multiple layers within a plot or beneath a point contact can result in an inaccurate representation of an intertidal population. For example, when taking cover data for a quadrat, if only the fronds of a rockweed overlying a dense group of barnacles are included in cover estimates, then the underlying barnacles will be erroneously missing from the dataset. One method of handling layering is to record all taxa by making separate determinations of overstory and understory layers (Littler and Littler 1985; Raffaelli and Hawkins 1996). Layering also can be accommodated in the dataset by recording the sequence in which each taxon is encountered. Because of layering, it is very common for individual plots or sampled intertidal habitats to support more than 100% cover, particularly in layered seaweed-dominated intertidal communities and mussel beds.

There are three basic methods for obtaining estimates of the percentage cover of rocky intertidal populations: (1) scanning plots visually, (2) determining the number of point intercepts or contacts along a line or within an area, and (3) tracing the silhouettes or photographic images of organisms by planimetry or with image-analysis software. In the last method, actual metric areas are then converted into percentages to obtain percentage cover data. On rare occasions, the cover of a species may be calculated by a formula based on measurements of selected dimensions, but this method is only applicable to species with regular or proportionate shapes and is not discussed further. The first two methods are commonly employed directly in the field, while all three methods can be used in the laboratory working from 35-mm slides, photographic prints, digital images, or video film.

Field Methods: Visual Scanning. Visual scanning is a commonly used field method that relies on an investigator's ability to accurately estimate the percentage of the planar surface area of a plot covered by the species or material of interest. Often, investigators divide plots into subsections (fig. 6.8) to break cover estimates down into smaller, more tractable units (e.g., sections representing 10% of the quadrat area) to facilitate estimates and improve accuracy (Dethier et al. 1993). When employed by an experienced observer, visual cover estimates can be completed rapidly in the field, even under harsh conditions, and result in an accurate taxonomic inventory of plot contents. For these reasons, visual scanning may be the preferred assessment method if a major goal of the study is to determine species richness.

Despite its advantages, the visual scanning method generally is considered less accurate than other, less subjective methods of determining cover, particularly when plot contents are complex (but see Dethier et al. 1993). Because investigators differ in their levels of experience and their inherent ability to visually group together taxa of different sizes, shapes, and distributions, visual cover estimates have a high potential for observer error.

Figure 6.8. Investigators performing cover estimates in plots divided into subsections. The practice of dividing plots into subsections is frequently used to facilitate visual estimates.

Moreover, errors produced by visual scanning are unknown because they are largely a property of the investigator and his or her taxonomic and visual abilities. For inexperienced observers the absolute error can be high, particularly when making estimates of small, spatially scattered taxa (e.g., chthamaloid barnacles) or species contributing low cover (e.g., 5% to 10%). Because different observers often arrive at different cover estimates in the same circumstances, visual scanning also is the least precise or repeatable of the methods commonly used in intertidal assessments (Rivas 1997). Accordingly, visual scanning may not be the method of choice in studies where multiple investigators repeatedly collect data from one or more sites, or in long-term monitoring programs where cover is sampled over a time span likely to transcend the participation of individual observers. Finally, most sampling programs require large numbers of replicates. The repetitive use of visual scanning can be mentally demanding, particularly under conditions of high biological complexity. For these reasons, errors in visually estimating cover can increase as the investigator tires during the later portions of an intensive assessment period.

To reduce field time and simplify sampling, some investigators have adopted the use of gridded quadrats and record species present only within individual grids or subsections in making abundance estimates. The data collected then actually constitute the percentage or frequency of grids within which a species occurs and are not actually cover values. Conversion to cover requires assigning each grid occurrence a cover value equal to the percentage of the quadrat it represents and then summing these values for the quadrat. This approach might work to generate reasonable cover estimates for large, abundant species, particularly if grids are numerous, but a likely outcome will be unacceptably high sampling errors, particularly for smaller-bodied species distributed widely throughout plots. Unless each individual subsection comprises a very small proportion of the total plot area, this procedure is not recommended except for surveys where only semiquantitative data are required. If fully quantitative visual cover estimates are believed to be impractical, data of higher value and greater utility easily can be obtained with similar expenditure of field time by using gridded intersections as point contacts and listing species not contacted by points as simply present in quadrats. For the latter, rare species, an arbitrary cover value can then be assigned that is equal to a value less than that possible when a species in the quadrat is contacted by only a single point.

Field Methods: Point Contacts. Besides visual scanning, the most commonly used field procedure for estimating the percentage cover of intertidal organisms is the point-intercept or point-contact method. Terrestrial plant ecologists have long used point-based procedures, and much is known about the advantages and disadvantages of point sampling terrestrial vegetation (Greig-Smith 1983). When using point-based methods, the taxon or taxa subtended by each point dropped in the sampling area must be determined and recorded by the investigator (fig. 6.9). Percentage cover of a species is then calculated by dividing the number of its point contacts by the total number of points distributed within the sampled area or plot. Various means can be used to provide point fields using this procedure (Kingsford and Battershill 1998). For example, marks (points) can be made on acetate, plastic sheets, or Plexiglas (=Perspex) boards (e.g., Dungan 1986; Meese and Tomich 1992), rods can be inserted through Plexiglas plates (e.g., Foster et al. 1991), knots can be tied randomly in string attached to a metal bar (Kingsford and Battershill 1998), and intersecting grid lines can be affixed to plot frames (Kennelly 1987) or provided as plastic mesh (Underwood et al. 1991). The use of a double layer for fixing or sighting points can reduce parallax errors (Hawkins and Jones 1992; Raffaelli and Hawkins 1996).

Figure 6.9. Field researchers making cover estimates using a point contact procedure. Stainless-steel rods are dropped vertically through holes in a leveled Plexiglas (Perspex) plate used to position rods.

A potential source of error in point sampling stems from the actual diameter of the pin or point used to determine contacts (Greig-Smith 1983). Although theoretically the contacts in point sampling are based on true points, actual sampling is performed using needles, pins, rods, cross-wires, grids of line or mesh, knots in string, laser light projections, or other objects that only approximate true points. These objects have a finite diameter and contact an area greater in size than a true point. An effect of using points with a finite diameter is that the percentage cover is overestimated, particularly when sampling smaller, finely branched plants (Goodall 1952).

Points can be distributed randomly or systematically in regular rows and columns. The relative merits of using random versus regularly distributed point fields have not been determined for sampling intertidal organisms (Dethier et al. 1993). For terrestrial plants, however, systematically arranged points yield cover values similar to those obtained with random point arrays if (1) the distance between points exceeds the size of the individuals or plant clusters, and (2) spacing does not correspond with a repetitive pattern in the vegetation (Tidmarsh and Havenga

1955). If points are randomized, a stratified approach should be used in most cases to ensure that points are dispersed over the entire area being sampled. This is accomplished in the field by dividing the sampling apparatus into sections (e.g., quarters) and then randomly selecting points from arrays arranged in regular rows and columns. This procedure establishes minimum distances between points, ensures that points are dispersed over the entire plot area, and enables the observer to easily keep track of assessed points when performing fieldwork.

A known advantage of point-based sampling procedures is that cover estimates become more accurate with increasing point densities (Greig-Smith 1983). This benefit is obtained, however, at the expense of the greater time required to process an individual sample. For this reason, the investigator should select a point density that yields an ecologically acceptable margin of error and a point field that can be processed within the time constraints of the sampling program. Hence, the number of points to be sampled should probably be determined from a pilot study. For most intertidal work, however, sampling has been based on history and tradition (Andrew and Mapstone 1987) instead of rigorous evaluation, and the densities used in point-contact sampling have frequently been selected from previous studies. Although Raffaelli and Hawkins (1996) suggest that 30 to 50 points per plot is appropriate for most studies, the majority of workers appear to use point fields ranging from 50 to 100 points (table 6.3). With 50 to 100 points per plot, the minimum point-generated cover value for a taxon will lie between 2% and 1%, the case when a single point hits a taxon. Gonor and Kemp (1978) provide a brief discussion of procedures for determining the number of points required to generate cover data and Kingsford and Battershill (1998) discuss how binomial sampling theory can inform selection of a suitable point field density.

The very high point densities required for accurately estimating the cover of species found in low abundance (<5% cover) are rarely, if ever, attained in field sampling programs because of sampling-effort limitations. Relying exclusively on point contacts to quantify species presence and cover not only provides poor estimates for species that occur in low abundance but often completely excludes rare species from plot assessments (e.g., Carter and Rusanowski 1978; Meese and Tomich 1992; Dethier et al. 1993; Rivas 1997). Many species commonly occur in low abundance in temperate rocky intertidal systems and, as a consequence, will be missed entirely by most point fields. As a result, taxonomic richness is likely to be underestimated in exclusively point-based sampling programs, as will diversity measures that are highly sensitive to species number. For example, Dethier et al. (1993) observed that a point density of 1 point per 50 cm^2 of primary substratum failed to include 19% of the

TABLE 6.3. Point Densities Used in Selected Intertidal Sampling Programs

Number of Points	Plot Area (m^2)	Reference(s)
49	0.06	Meese and Tomich (1992)
50	0.25	Dethier et al. (1993)
75	0.15	Denis and Murray (unpublished data)
100	0.04	Benedetti-Cecchi et al. (1996)
100	0.15	Foster et al. (1991), Rivas (1997), Sapper (1998).
150	0.15	Littler and Littler (1985)

species present, with all omitted taxa contributing less than 2% cover. Consequently, it is impractical in most cases to attain the sampling effort needed to meaningfully measure species richness or detect spatial or temporal differences in the cover of rarer species using only point-contact sampling methods.

Dethier et al. (1993) have argued that cover estimates made using point-contact procedures require more sampling time than estimates using visual scanning. This might not be true, however, when sampling programs require observers to process many samples in a single field day or when plot contents are multilayered and complex. Under these conditions, determining the species hit by points results in much less fatigue and accelerates the processing of samples compared with the visual conversion of shapes and patterns into cover values. Studies by Rivas (1997), who used simulated species made of paper and cardboard, and field experience by S. N. Murray support the premise that point-contact methods are usually less time-consuming than visual scanning when processing many samples. Moreover, the time required to process a sample is less influenced by biological complexity when point-contact procedures are employed.

Photographic Sampling

Density and cover estimates also can be made in the laboratory from photographic methods that obtain images in the field using 35-mm film, video, or digital cameras. Traditionally, this has been accomplished using 35-mm cameras and color (64 or 100 ASA slide) or black and white film, although portable Hi 8-mm and, more recently, digital format camcorders and digital still cameras have become available. Organisms can be counted and the same visual and point-contact procedures for estimating cover in the field can be used on the resulting slides, prints, video film, or digital images with most of the same advantages and disadvantages. Photographs or video and digital images also can be used in the laboratory to determine the areas of organisms using a handheld planimeter or with the aid of a computer using digitized tracings and image-analysis software. Photographic sampling methods offer several advantages over methods that depend exclusively on *in situ* field assessments but also have significant disadvantages.

Intertidal studies are almost exclusively performed during periods of low tide so access to the shore is limited and investigators encounter field-time constraints that restrict the number of samples that can be collected. Photographic sampling can be completed much more rapidly than full field assessments and can increase the number of samples taken per unit field time. Because plot-to-plot variation is usually high in intertidal habitats, a large number of samples is often needed to obtain the statistical power necessary to detect ecologically meaningful differences in species abundances. Photographic sampling also offers the advantage that an essentially permanent record of the contents of each plot can be archived and later independently analyzed by other investigators or used to visually demonstrate changes in species composition and abundances, even to nonscientific audiences. Additionally, *in situ* quantification errors resulting from investigator mistakes or fatigue, inclement weather, or dangerous surf can be reduced, and parallax problems resulting in edge-effect errors can be eliminated by photographic sampling (Gonor and Kemp 1978).

Unfortunately, laboratory-based sampling methods that rely on photographs, video records, or digital images also have significant disadvantages. The most important of these is the limits of resolution of the film, tape, or digital media, which make it difficult to locate small, "difficult-to-see" taxa or to accurately distinguish morphologically similar species (Gonor and Kemp 1978; Foster et al. 1991). Photographic sampling procedures also present problems when applied to highly layered intertidal communities (Gonor and Kemp 1978; Foster et al. 1991; Meese and Tomich 1992; Leonard and Clark 1993) because the captured images are only two-dimensional views and the entire three-dimensional plot contents cannot be inspected as they can in the field. Thus, photographic sampling can be severely compromised in dense seaweed or mussel communities, where it is impossible to easily move aside upper layers and photograph underlying organisms. Photographic techniques work best when sampling simple communities consisting of largely two-dimensional populations such as small barnacles, crustose algae, or only the overstory fronds of large, canopy-forming seaweeds. Unfortunately, highly layered and often seaweed-dominated communities are common throughout the middle and lower intertidal zones of most temperate coasts. For these reasons, Foster et al. (1991) emphasize the need to substantiate identifications and quantification of abundance through on-site fieldwork when using photographic sampling methods. Photographic sampling procedures also place constraints on the size (and shape) of sample plots; these must be small enough to be photographed at a scale where plot contents can be identified and of a shape that conveniently fits within the film frame.

Thirty-five-millimeter Photographic Sampling. Thirty-five-millimeter single-lens-reflex (SLR) cameras have long been used to assess species abundances in intertidal sampling programs (Littler and Littler 1985). The interested reader can find a discussion of the use of 35-mm cameras in marine sampling programs in George (1980). For most applications, photographs are taken of framed quadrats, in which case the known quadrat dimensions provide a built-in linear scale. Rectangular quadrats often are preferred for photographic sampling because their dimensions match those of the 35-mm film format (fig. 6.10). Nonreflective Plexiglas (=Perspex) can be fixed to the exterior of the quadrat frame and written on with a wax pencil or lumber crayon to record quadrat labels such as date, site, and quadrat number on the film (Littler and Littler 1985).

A wide variety of films, cameras, and lenses is available at a range of prices. Usually, 35-mm color transparency (slide) film (ASA 64 or 100) is preferred because color offers the best opportunity to distinguish intertidal organisms that are variable in coloration and contrast, and because the ability to project or digitize transparencies facilitates laboratory assessment. Infrared color film can be advantageous for photographically sampling certain seaweeds, particularly, thick cyanobacterial (=blue-green algal) and diatom films that occur on intertidal rock surfaces (Littler and Littler 1985). Generally, a 35-mm SLR camera for intertidal work will be equipped with a good-quality lens and be resistant to damp, salty conditions. The choice of lens will depend on the size of the area to be photographed, but for many applications a zoom lens (e.g., 35 to 70 or 80 mm) may be preferred to facilitate framing and focusing. However, for most cameras and lens systems, a fixed 35-mm lens will provide greater photographic resolution. An easily accessible shoe or PC port that allows the use of a battery-powered, electronic strobe is essential because sampling often must be performed in poor light—such as at night, in the early morning, or late afternoon. The additional light enables the use of high lens *f*-stops (*f*8 or greater) to maximize the depth of field in photographs. Strobe lighting also can remove shadows and reduce contrast during well-lit midmorning or afternoon hours. When high contrast between subjects and shadows is produced by natural lighting conditions,

Figure 6.10. Investigators using a handheld 35-mm camera equipped with an electronic strobe to photograph the contents of a field plot. The photographer has taken a position directly above the plot so as to photograph its contents at an angle as close as possible to 90° to the plane of the substratum. Usually, a second person is required to provide assistance in positioning and securing plots on shores where the topography is heterogeneous and irregular.

photographic quality can be improved greatly by using strobe lighting in conjunction with an umbrella or other object to uniformly shade the sample area. Adjustments in film exposure often will be required to achieve good results when the photographer moves between intertidal communities. For example, barnacle and rock-dominated upper shore plots generally require less exposure time (accomplished by an increase of one or two *f*-stops) than plots harboring darker-colored mussels. Strobes should be mounted on a bracket to the side of the camera to avoid reflection from wet rocks and seaweeds. A polarizing filter can be used to reduce glare, but when one is used, care must be taken to ensure that the filter is set to provide polarization and that appropriate adjustments have been made in setting the *f*-stop or exposure time.

A significant disadvantage of relying on 35-mm photographs is that it is difficult or impossible to later compensate for poor images, and even experienced investigators can have trouble obtaining consistently good exposures under the full range of field conditions. Moreover, because 35-mm film must be laboratory processed, the quality of photographs cannot be checked before leaving the field. To ensure that properly exposed photographic samples are obtained, one or two additional photographs can be bracketed around the selected settings by using different lens *f*-stops or exposure times. Film is much less costly than the cost (time and effort) required for field sampling, so bracketing exposures to ensure usable photographic samples is always a good idea. Most of these disadvantages can be overcome through the use of digital photographs with cameras that allow the investigator to preview images in the field.

The enclosure of 35-mm SLR cameras and strobes in waterproofed protective cases also might be necessary under rigorous field conditions where unprotected cameras and strobes are exposed to precipitation, ocean spray, or the possibility of accidental submergence. However, enclosure in protective cases requires an added expense and, because cases often are bulky, can make 35-mm SLR cameras inconvenient to use. To overcome these problems, most researchers rely on underwater cameras (e.g., Nikonos 35-mm series) for sampling under wet field conditions. Unfortunately, many underwater cameras, including reasonably priced Nikonos models, are range-finder cameras that do not frame or focus the subject directly through the lens. This increases the probability of framing and focusing errors when range-finder cameras are handheld and used by inexperienced photographers. Framing and focusing problems with underwater cameras can be overcome by mounting the camera on a fixed stand, made, for example, from PVC pipe, that frames the sample and then setting the lens at a predetermined focal distance from the subject; single or paired strobes can be mounted at preset lateral positions on the stand to provide uniform and angled lighting (fig. 6.11). Although fixed camera stands are convenient under most field conditions, they take field time to assemble and can be difficult or cumbersome to use when plots occur on vertically positioned or irregularly angled substrata.

To partially overcome layering problems in seaweed-dominated communities, both overstory and understory photographs of the same plot can be taken and then separately scored in the laboratory (Littler and

Figure 6.11. Photographer standing adjacent to a Nikonos camera and an electronic strobe mounted to a fixed stand made of PVC pipe. The base of the stand has been configured to represent the plot borders. All photographs are taken at the same fixed angle and distance with respect to the plot surface.

Littler 1985). In performing this procedure, a single photograph of the overlying seaweed canopies is first taken and then the understory layer is photographed after canopy fronds are nondestructively combed aside. Nonetheless, it is almost always impossible to clearly view all plot contents down to the primary substratum using photographs. To better interpret plot contents and provide information on "difficult-to-see" taxa, 35-mm photographs can be supplemented with field estimates of species cover and notes describing the locations of species within plots (Littler and Littler 1985). By constructing data sheets with rows and columns that match plot subsections, crude maps of species locations within plots can be obtained simultaneously with estimates of subsection cover. Although field notes can add important information, if notes are detailed, this effort can be time consuming and defeat one of the principal advantages of photographic sampling: reduction in the amount of field time required to obtain samples.

Video Sampling. A few years ago, Whorff and Griffing (1992) introduced a video procedure for sampling rocky intertidal populations. This method employed VHS Camcorders to obtain video images from which quantitative abundance data were later extracted in the laboratory. Various video-based techniques have been used for some time for performing benthic surveys (Harris 1980; Holme 1984, 1985; Thouzeau and Hily 1986; Leonard and Clark 1993), especially from submersibles (e.g., Potts et al. 1987), and sampling fish in the water column (Kingsford 1998). Despite its potential (Whorff and Griffing 1992; Leonard and Clark 1993), video sampling has not been used widely in rocky intertidal research, and most photographic methods for sampling rocky intertidal populations have relied on 35-mm cameras and film. Nevertheless, video technology has improved rapidly in recent years, a trend that clearly will continue in the future. Highly portable, battery-operated Camcorders, lighting accessories, and video playing and recording systems are now readily available at relatively low cost. Hi 8-mm video film and digital video represent significant advances in quality over both VHS and SVHS video media and can yield sharp and informative images.

Video methods provide many of the same advantages and disadvantages as those discussed previously for 35-mm photographic sampling. However, the greatest concern when using affordable video in place of 35-mm film is image quality. For most applications, neither digital nor video film Camcorder media are yet able to offer comparable resolution to 35-mm film. For many purposes, the reduction in resolution obtained with video film or digital media may not be crucial, and samples can be quantified at levels comparable to those attainable with 35-mm film. However, the higher image quality of 35-mm film may be a requirement for many other uses. The use of digital still cameras in place of 35-mm SLR film cameras has the same disadvantage: images will still generally be of lower resolution than those captured on 35-mm, although with each passing year this gap is being closed with improvements in digital technology, and many experienced field-workers are now moving to the use of digital cameras for field photography.

In contrast to its disadvantages, video can also offer advantages and can overcome some of the problems associated with 35-mm photographic sampling. Because adjustments in filming angles, film exposure, and lighting can be easily done in the field, it is possible to quickly take multiple frames of the same area to increase the likelihood of acquiring images of needed quality. Additionally, Camcorders provide the additional advantage of being able to zoom in and out while filming. Zooming in to produce close-ups of small plot sections can help overcome difficulties in identifying "difficult-to-see" taxa and increase the amount of information obtained beneath seaweed canopies. Besides filming whole plots, Rivas (1997) also filmed smaller areas nested within each larger plot to provide supplemental close-up video footage used to help identify difficult-to-see taxa and to quantify the understory contents of full plots (fig. 6.12). Plots (0.3 × 0.5 m) were divided into quarters, which were in turn divided into two rows and three columns to produce six subsections. After filming each plot quarter, the smaller, understory species in the six subsections were filmed sequentially by zooming-in the lens while holding the camera steady and perpendicular to the substratum.

Laboratory Assessment of Photographic Samples. Visual observations of photographic records can be used in the laboratory to list species, count individuals, or visually estimate percentage cover. For most layered intertidal communities with well-developed seaweed canopies, small or cryptic species and low-abundance taxa are often difficult to identify or count based only on information contained in photographic samples. Hence, photographic records rarely can be used alone to obtain full taxonomic inventories or to determine the densities of most species. On the other hand, 35-mm color slides often have been used for laboratory determinations of the cover of intertidal species using visual (e.g., Lubchenco and Menge 1978) or point-contact (Bohnsack 1979; Littler and Littler 1985; Foster et al. 1991) methodologies that also can be transferred readily to a video-based or digital platform.

Usually, 35-mm film is processed to produce color slides or transparencies that can be projected onto a screen or viewed directly using a dissecting microscope. Alternatively, 35-mm film can be scanned to produce digital images for enhancement or to enable quantification using computer software. Slides can be projected directly onto a point grid or a sheet of acetate and percentage cover estimates made using point-contact or other methods (fig. 6.13); alternatively, digital images can be viewed on a computer screen and scored using acetate sheets or software displaying point fields. For example, Littler and Littler (1985) describe a procedure where 35-mm color slides are projected and focused onto a sheet of fine-grained, white paper containing a grid of points arranged at 2-cm

Figure 6.12. Videographer filming plot contents in the field with a Camcorder equipped with a battery-operated light source. Filming procedures include taking plot views at different scales ranging from full plot photos to macrophotos of small plot sections.

intervals. The organisms subtended by each point are then identified and tallied by the investigator with help from written maps and data sheets or tape-recorded field notes. Alternatively, the intersections of grid lines in a standard ocular graticle (e.g., 10 × 10) can be used as points to score 35-mm slides or photographic prints viewed with a dissecting microscope.

Although labor-intensive, the photographic sampling program described by Littler and Littler (1985) includes most of the attributes of both field and photographic data acquisition. In this procedure, cover was determined for the more abundant and easy-to-discriminate species in the laboratory based on point-contact scores obtained from 35-mm color slides. However, for the smaller, "difficult-to-see" taxa and understory species hidden in photographs, final cover values were based on field estimates because these could not be quantified with accuracy solely using photographs.

Similar methods can be used to process the contents of video samples. First the film is reviewed on a high-resolution monitor perhaps using a player/recorder equipped with an editing box or computer software that allows frame-by-frame scrolling. Video frames containing good images of

Figure 6.13. Slides being scored in the laboratory using field notes. A 35-mm color slide of a field plot has been projected onto a piece of paper containing a grid of points. The grid is taped to a glass panel and the slide is projected from a position on the opposite side of the panel to eliminate shadowing.

each quadrat can then be selected and frozen on the monitor or digitized from film using a computer equipped with a video capture card. Digital photographs taken with still or video cameras eliminate the need to use a capture card to digitize film. Captured digital images can be labeled and stored on disks or other media in various formats (e.g., as TIFF files) until analyzed. Rivas (1997) describes a procedure where video frames were captured and stored as TIFF files, then opened in Adobe Photoshop, where TIFF images were framed and enhanced, if necessary. Images were then transferred to NIH Image, an image-analysis program that can be downloaded from the NIH Web site (http://rsb.info.nih.gov/nih-image). Rivas then employed a custom-designed macro program, embedded in NIH Image, to display a prescribed number of randomly dispersed points over the image surface. The points generated by the macro were used to determine percent cover of filmed species using the point-contact procedures described previously. In lieu of using NIH Image or a comparable image-analysis program, a sheet of clear acetate containing a point grid can be superimposed on the high-resolution monitor or the computer screen to process the video samples. Either field notes or close-up video of plot sections can be used to facilitate identification of taxa. If video close-ups of plots are available, these can be played back simultaneously on an adjacent monitor to identify small "difficult-to-see" species and to obtain other views of species lying beneath layered seaweed fronds.

In addition to the above procedures, photographs or video records can be used to make area determinations of crustose or thick, coarse taxa by tracing perimeters with a planimeter. Alternatively, 35-mm film or video images can be digitized and tracings can be performed using computer-based image-analysis programs such as NIH Image. However, planimeter or digital-based estimates of occupied areas are really only feasible for larger, coarsely branched seaweeds or crustose algae and animals because of difficulties in accurately tracing finely branched or small, irregularly shaped forms. Recently, Pech et al. (2004) described a procedure for making abundance estimates of rocky intertidal invertebrates from digital photographs using digital-image analysis.

SUMMARY

Abundance data are routinely collected to describe the status and dynamics of rocky intertidal populations and communities and, therefore, form the foundation of most rocky intertidal sampling programs, including monitoring and impact studies. How to collect abundance data becomes an important sampling consideration, and as with decisions on sampling design and sampling units (see chapters 4 and 5), the selected procedures must be carefully matched with the goals of the sampling program.

Qualitative representations of species abundances can be made using arbitrary categories or numerical scales during quick surveys, but these data will be of limited comparative value, particularly in time-series monitoring programs, because of difficulties in standardizing observer estimates. Point sampling along transect lines can be used for surveys of space-occupying macrophytes and sessile invertebrates when more time is available. This is an efficient and repeatable procedure that reduces subjectivity, provides better precision, and, for most applications, is preferable to strictly qualitative procedures for most purposes. Although other plotless procedures besides line and point techniques are available, these are used less often than plot-based methods in most rocky intertidal studies.

Abundance data are usually collected in the field by counting individual organisms or by estimating percentage cover visually or through the use of point-contact methods. Usually, these data can be accurately obtained using nondestructive sampling procedures, except in certain highly layered communities such as mussel beds. Density data can be obtained only for species with discrete individuals, and cannot be consistently collected for most seaweeds and many colonial invertebrates. For these species, and also for abundant, sessile invertebrates such as barnacles, percentage cover is commonly used to express abundance in terms of space occupancy. Cover values for mobile invertebrates, most of which occur at much lower densities than barnacles, have little ecological meaning and are difficult to interpret. Because of the inability to count most seaweeds and the ecological meaning of cover values for mobile invertebrates, most community-level studies do not rely on one abundance parameter but instead rely on cover data for seaweeds (and often sessile invertebrates) and density data for mobile invertebrates.

Edge effects present problems in plot-based sampling and can become an important source of error when estimating species abundances. This problem can be enhanced by observer parallax and requires that efforts are made to reduce bias during the quantification of plot contents. Layering at larger (e.g., algal canopies) and smaller (e.g., epibiota) scales also present significant sampling problems and often result in >100% plot cover, particularly in seaweed-dominated intertidal communities and mussel beds.

Field data collection provides the best opportunity for recording the presence of species in sampling units and also for quantifying the abundances of "difficult-to-see," morphologically similar, and rare taxa. Visual cover estimates collected by an experienced observer should result in an accurate inventory of plot contents and can be completed rapidly. However, visual cover estimates are often less accurate and are less repeatable than more objective point-contact methods; visual estimates also have unknown sampling errors that are largely the property of the taxonomic and visual abilities of the observer. For these reasons, visual methods should not be used in long-term monitoring programs where multiple observers collect data at one or more sites over a lengthy time span. Gridded plots, where cover estimates are derived from the presence or absence of species within grid units, might be satisfactory for sampling abundant species but generally are not recommended where the goal is to determine statistical changes in species abundance. Point-contact procedures have long been used in sampling terrestrial vegetation and also have been widely used in rocky intertidal sampling programs. Point-based procedures are easy to apply and become more accurate with increasing point densities, meaning that an investigator can adjust the number of points to obtain the desired margin of sampling error. Point methods also are more repeatable and can be less tiring and time-consuming than visual procedures, particularly where biological complexity is high. Hence, point-contact approaches are preferred over visual cover estimates in monitoring programs or other time-series sampling studies where it is important to reduce sampling error and where multiple observers participate in data collection.

Photographic sampling procedures, using 35-mm, video, or digital cameras, reduce the field time required per sample, eliminate parallax problems, and yield permanent records that can be archived or inspected by other investigators. These are valuable attributes for long-term monitoring programs. However, the ability to count individuals, to discriminate among morphologically similar species, and to resolve difficult-to-see taxa present significant disadvantages, particularly in highly layered intertidal communities. Methods for extracting data from photographic samples in the laboratory are well established but generally are as time-consuming as collection of the same data in the field. However, laboratory analyses can be completed over a much longer time schedule, under comfortable conditions, and at the convenience of the investigator. Photographic samples also are excellent tools for reconstructing time-series changes and, therefore, are valuable components of any long-term intertidal monitoring program, either for data collection or for providing archivable visual records of sample contents.

LITERATURE CITED

Ambrose, R. F., P. T. Raimondi, and J. M. Engle. 1992. Final study plan for inventory of intertidal resources in Santa Barbara County. Unpublished Report to Minerals Management Service, Pacific OCS Region.

Andrew, N. L., and B. D. Mapstone. 1987. Sampling and the description of spatial pattern in marine ecology. *Oceanogr. Mar. Biol. Annu. Rev.* 25:39–90.

Baker, J. M. and J. H. Crothers. 1987. Intertidal rock. In *Biological surveys of estuaries and coasts,* ed. J. M. Baker and W. J. Wolff, 157–97. Cambridge: Cambridge Univ. Press.

Benedetti-Cecchi, L., L. Airoldi, M. Abbiati, and F. Cinelli. 1996. Estimating the abundance of benthic invertebrates: a comparison of procedures and variability between observers. *Mar. Ecol. Progr. Ser.* 138:93–101.

Bohnsack, J. A. 1979. Photographic quantitative sampling of hard-bottom benthic communities. *Bull. Mar. Sci.* 29:242–52.

Boudouresque, C. F. 1969. Une nouvelle méthode d'analyse phytosociologique et son utilisation pour l'étude des phytocoenoses marines benthiques. *Téthys* 1:529–34.

———. 1971. Méthodes d'étude qualitative et quantitative du benthos (en particulier du phytobenthos). *Téthys* 3:79–104.

Braun-Blanquet, J. 1927. *Pflanzensoziologie.* Wien, Germany: Springer-Verlag.

Brower, J. E., J. H. Zar, and C. N. von Ende. 1998. *Field and laboratory methods for general ecology.* 4th ed. Boston: WCB/McGraw–Hill.

Carter, J. W., and P. C. Rusanowski. 1978. A point contact sampling methodology for marine ecological surveys with comparison to visual estimates. In *Proceedings of the National Conference on the Quality of Assurance of Environmental Measurements, Nov. 27–29, 1978,* 65–73. Denver, CO: Hazardous Materials Control Research Institute Information Transfer.

Creese, R. G., and M. J. Kingsford. 1998. Organisms of reef and soft substrata intertidal environments. In M. Kingsford and C. Battershill, 1998, 167–93.

Crisp, D. J., and A. J. Southward. 1958. The distribution of intertidal organisms along the coasts of the English Channel. *J. Mar Biol. Assoc. U.K.* 37:1031–48.

Dawson, E. Y. 1959. A primary report on the benthic marine flora of southern California. *In* An oceanographic and biological survey of the continental shelf area of Southern California. Publs. *Calif. St. Water. Pollut. Control. Bd.* 20:169–264.

———. 1965. Intertidal algae. *In* An oceanographic and biological survey of the Southern California mainland shelf. Publs. *Calif. St. Water. Qual. Control. Bd.* 27:220–31, 351–438.

Dethier, M. N., E. S. Graham, S. Cohen, and L. M. Tear. 1993. Visual versus random-point percentage cover estimations: 'objective' is not always better. *Mar. Ecol. Progr. Ser.* 96:93–100.

Dungan, M. L. 1986. Three-way interaction: barnacles, limpets, and algae in a Sonoran Desert rocky intertidal zone. *Am. Nat.* 127:292–316.

Foster, M. S., C. Harrold, and D. D. Hardin. 1991. Point vs. photo quadrat estimates of the cover of sessile marine organisms. *J. Exp. Mar. Biol. Ecol.* 146: 193–203.

George, J. D. 1980. Photography as a marine biological research tool. In *the shore environment, Vol. 1. Methods*, ed. J. H. Price, D. E. G. Irvine, and W. F. Farnham, 45–115. London: Academic Press.

Gonor, J. J., and P. F. Kemp. 1978. Procedures for quantitative ecological assessments in intertidal environments. U.S. Environmental Protection Agency Report EPA-600/3-78-087.

Goodall, D. W. 1952. Some considerations in the use of point quadrats for the analysis of vegetation. *Aust. J. Sci. Res. Ser. B* 5:1–41.

Greig-Smith, P. 1983. *Quantitative plant ecology*. 3rd ed. Berkeley: Univ. of California Press.

Harris, R. J. 1980. Improving the design of underwater TV cameras. Intern. *Underwat. Syst. Designs* 2:7–11.

Hawkins, S. J., and H. D. Jones. 1992. *Rocky shores* (Marine Conservation Society, Marine Field Course Guide 1). London: Immel.

Hayek, L.-A., and M. A. Buzas. 1997. *Surveying natural populations*. New York: Columbia Univ. Press.

Holme, N.A. 1984. Photography and television. In *Methods for the study of marine benthos,* ed. N. A. Holme and A. D. McIntyre, 66–98. Oxford: Blackwell Scientific.

———. 1985. Use of photographic and television cameras on the continental shelf. In Underwater photography and television for scientists, ed. J. D. George, G. I. Lythgoe, and J. N. Lythgoe, 88–99. Oxford: Clarendon Press.

Kennelly, S. J. 1987. Inhibition of kelp recruitment by turfing algae and consequences for an Australian kelp community. *J. Exp. Mar. Biol. Ecol.* 112: 49–60.

Kershaw, K. A. 1973. *Quantitative and dynamic plant ecology*, 2nd ed. New York: American Elsevier.

Kingsford, M. J. 1998. Reef fishes. In *Studying temperate marine environments. A handbook for ecologists, ed.* M. J. Kingsford and C. N. Battershill, 132–66. Christchurch, New Zealand: University Press.

Kingsford, M. J., and C. N. Battershill. 1998. Subtidal habitats and benthic organisms of rocky reefs. In *Studying temperate marine environments. A handbook for ecologists,* ed. M. J. Kingsford and C. N. Battershill, 84–114. Christchurch, New Zealand: Canterbury University Press.

Krebs, C. J. 1989. *Ecological methodology*. New York: Harper & Row.

Leonard, G. H., and R. P. Clark. 1993. Point quadrat versus video transect estimates of the cover of benthic red algae. *Mar. Ecol. Progr. Ser.* 101:203–8.

Littler, M. M., and D. S. Littler. 1985. Nondestructive sampling. In *Handbook of phycological methods. Ecological field methods: Macroalgae,* ed. M. M. Littler and D. S. Littler, 161–75. Cambridge: Cambridge Univ. Press.

Loya, Y. 1978. Plotless and transect methods. In *Coral reefs: research methods,* ed. D. R. Stoddart and R. E. Johannes, 197–217. Paris: UNESCO.

Lubchenco, J., and B. A. Menge. 1978. Community development and persistence in a low rocky intertidal zone. *Ecol. Monogr.* 48:67–94.

Maggs, C. A., and D. P. Cheney. 1990. Competition studies of marine macroalgae in laboratory culture. *J. Phycol.* 26:18–24.

Magurran, A. E. 1988. *Ecological diversity and its measurement.* Princeton, NJ: Princeton Univ. Press.

Meese, R. J., and P. A. Tomich. 1992. Dots on the rocks: a comparison of percentage cover estimation methods. *J. Exp. Mar. Biol. Ecol.* 165: 59–73.

Mueller-Dombois, D., and H. Ellenberg. 1974. *Aims and methods of vegetation ecology.* New York: John Wiley & Sons.

Murray, S. N., and P. S. Dixon. 1992. The Rhodophyta: some aspects of their biology. III. *Oceanogr. Mar. Biol. Annu. Rev.* 30:1–148.

Pech, D., A. R. Condal, E. Bourget, and P.-L. Ardisson. 2004. Abundance estimation of rocky shore invertebrates at small spatial scale by high-resolution digital photography and digital image analysis. *J. Exp. Mar. Biol. Ecol.* 299:185–99.

Pielou, E. C. 1984. The interpretation of ecological data. A primer on classification and ordination. New York: John Wiley & Sons.

Potts, G. W., J. W. Wood, and J. M. Edwards. 1987. Scuba-diver operated low-light level video system for use in underwater research and survey. *J. Mar. Biol. Assoc. U. K.* 67:299–306.

Raffaelli, D., and S. Hawkins. 1996. *Intertidal ecology.* London: Chapman and Hall.

Rivas, O. O. 1997. *Laboratory evaluation of a video method for sampling rocky intertidal populations.* M.A. thesis. Fullerton: California State Univ.

Sapper, S. A. 1998. *Variation in an intertidal subcanopy assemblage dominated by the rockweed* Pelvetia compressa (*Phaeophyceae, Fucales*). M.S. thesis. Fullerton: California State University.

Thouzeau, G., and C. Hily. 1986. A.Qua.R.E.V.E; une technique nouvelle d'echantillonnage quantitif de la macrofaune epibentique des fronds meubles. *Oceanol. Acta* 9:509–13.

Tidmarsh, C. E. M., and C. M. Havenga. 1955. The wheel-point method of survey and measurement of semi-open grasslands and karoo vegetation in South Africa. *Mem. Bot. Surv. S. Afr.* 29:1–49.

Trudgill, S. 1988. Integrated geomorphological and ecological studies on rocky shores in southern Britain. *Field Studies* 7:239–79.

Underwood, A. J., and M. G. Chapman. 1989. Experimental analyses of the influences of topography of the substratum on movements and density of an intertidal snail, *Littorina unifasciata. J. Exp. Mar. Biol. Ecol.* 134:175–96.

Underwood, A. J., M. J. Kingsford, and N. L. Andrew. 1991. Patterns in shallow subtidal marine assemblages along the coast of New South Wales. *Aust. J. Ecol.* 6:231–49.

Whorff, J. S., and L. Griffing. 1992. A video recording and analysis system used to sample intertidal communities. *J. Exp. Mar. Biol. Ecol.* 160:1–12.

Great Blue Heron searching for food at Crystal Cove State Park, in southern California.

CHAPTER 7

Quantifying Abundance

Biomass

Abundances of rocky intertidal organisms are generally expressed as numerical density or counts and as cover. Occasionally, however, the goals of a sampling program require that the biomass of rocky intertidal organisms be determined and used to express abundance. Biomass is defined as the mass of living organisms in a population at the time of sampling (Poole 1974). Biomass data usually are expressed as wet, dry, or ash-free dry weight per unit area, but also can be converted to calories, carbon, or other units. Generally, these conversions are made using mathematical relationships between biomass and the metric (e.g., calories) of interest. Previous discussions of procedures for sampling for biomass can be found in Gonor and Kemp (1978), Kanter (1978), Littler (1980), and DeWreede (1985). Brinkhuis (1985) also offers particularly good advice on how to treat algal samples during weight determination.

Biomass data represent the standing stock or the quantity of living material obtained at the time of sampling. Unlike density determinations, which require the ability of the investigators to distinguish individual organisms, biomass can be measured for all seaweeds and invertebrates. Collecting biomass data usually requires destructive sampling because organisms generally are removed from the habitat and returned to the laboratory to be weighed. Although as a rule sampling for biomass is destructive, the collected data can be of great value. For example, biomass data can be used as the common metric for reporting the abundances of all populations in a rocky intertidal community. As discussed in chapter 6, it is often impossible to distinguish discrete individuals when sampling most intertidal seaweeds and invertebrates such as sponges and colonial ascidians. Since investigators are generally unable to make counts of these populations, cover data usually are obtained to express the abundances of seaweeds and colonial invertebrates. In contrast, density, not cover, is the preferred metric for quantifying the abundances of mobile invertebrates. Consequently, biomass data can be used to express the abundances of all seaweeds and invertebrates in a community. This allows all sampled macroorganisms to contribute to community-level analyses dependent on diversity indexes or multivariate methods based on similarity matrices.

Compared with cover estimates, biomass data will generally better reflect the true abundances of most macrophytes and certain invertebrates in rocky intertidal communities. This is because cover is a two-dimensional (area-based) parameter that has limited ability to capture the quantity of plant or animal material that during high tide extends up from the substratum. Hence, biomass data generally best enable estimates to be made of resource availability between trophic levels. Biomass data (g m^{-2}) also can be used alone (e.g., Bellamy et al. 1973) or together with biomass-based measurements of photosynthesis (e.g., Littler and Murray 1974; Littler et al. 1979) to calculate the contributions of seaweed and seagrass populations to the community primary productivity. If productivity values are available, biomass data also can be used to calculate production-to-biomass ratios for a population or community. Biomass sampling can also increase the information obtained in biodiversity studies because inconspicuous species that might be difficult to see in the field can be encountered and inventoried during laboratory sorting. As a result, destructive biomass sampling usually results in more species determinations (and higher estimates of biodiversity) than nondestructive field sampling programs.

Using biomass data to estimate abundances of intertidal organisms also has several disadvantages (Gonor and Kemp 1978; DeWreede 1985). First, biomass sampling usually disturbs the study site because organisms almost always must be removed from the substratum and returned to the laboratory to be weighed. The destructive collection of biomass samples can affect future work, particularly in monitoring or other research programs that require repeated assessments of large, long-lived species with limited recruitment (Gonor and Kemp 1978). Although often difficult to detect, changes in the densities of opportunistic algae or mobile herbivores and predators following biomass harvesting can significantly influence community composition in areas adjacent to cleared plots. Hence, when repeated assessments of biomass are planned, the study area must be large enough to avoid reharvesting the same plots, and newly harvested sites must be spread at distances great enough to avoid the effects of prior sampling. These issues must be given careful consideration when designing long-term monitoring programs since there are no guidelines for arriving at "safe" distances between plots. Moreover, if the study area must be expanded to accommodate the demands of destructive sampling,

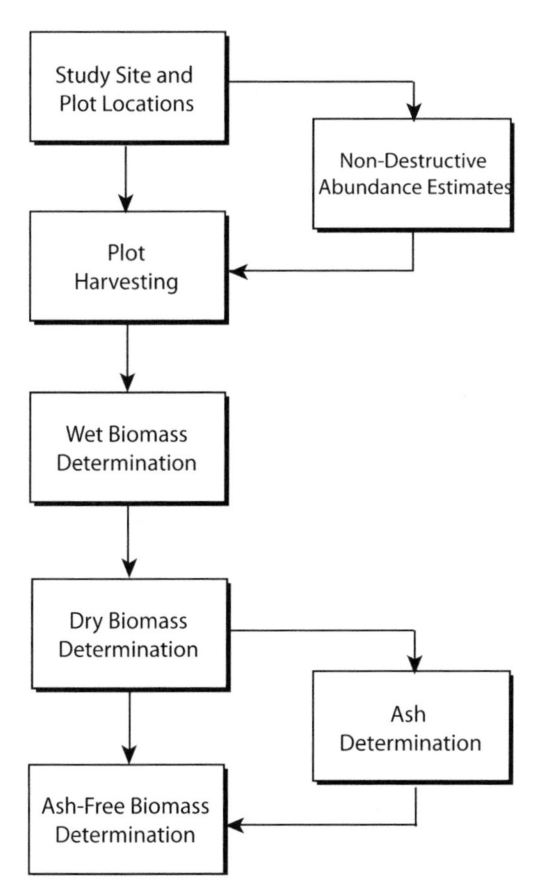

Figure 7.1. Decision tree for destructive biomass sampling.

then statistical comparisons of the collected data might be compromised by the environmental differences likely to occur as samples are distributed over more intertidal habitat (DeWreede 1985).

Besides altering community composition in the harvested plots, biomass determinations usually are very time-consuming and costly due to the need for extensive laboratory processing. For example, a single 0.3 × 0.5-m sample harvested in a dense, heavily epiphytized coralline algal turf can take 40 hr or more of laboratory work to sort seaweeds and invertebrates to species and perform wet and dry weight determinations. For this reason alone, biomass sampling is not suitable for low-cost sampling programs or survey work involving multiple sites.

Because of the destructive nature and high cost, biomass-sampling programs should be carefully planned and designed (figure 7.1). Careful consideration must be given to site selection (chapter 2) and choice of the specific area within the site to be delineated for study. Biomass estimates must be normalized by area for use as abundance data, requiring that plots or quadrats be used as sampling units. After selecting the plot size and shape, a strategy must be developed for locating samples within the study area. Considerations for distributing samples and for selecting plot sizes and numbers have been discussed by DeWreede (1985) and are addressed in chapters 4 and 5. Gonor and Kemp (1978) suggest that information should be obtained on the composition of each plot prior to extracting organisms. This can be done by taking 35-mm or digital photographs, making video records, or taking notes that describe species cover and by counting individuals and measuring a selected morphometric parameter to obtain size (e.g., maximum shell length). Complete nondestructive sampling for density or cover (see chapter 6) can be performed prior to removing organisms for biomass determination. If cover or density and size data have been taken prior to biomass determination, then biomass can be estimated from these data when implementing future nondestructive sampling programs. This can be accomplished, for example, by establishing mathematical relationships using the pairs of cover and biomass data gathered for each species from the harvested plots. However, such relationships require assumptions (e.g., minimal spatial and temporal variation) and should be used with caution. The strongest mathematical relationships should occur between cover and biomass for seaweed species with small fronds that generally do not show layering or extensive vertical growth and for largely two-dimensional invertebrates such as barnacles and sponges. For canopy-forming seaweeds, the best cover–biomass relationships should be achieved when the cover exceeds 5%–10% but is less than 80%–90%. This is because a larger, nondestructive, sampling error is expected when cover is low or approaches 100% when fronds are layered and lie on top of one another. Littler (1979a, 1979b) reported statistically significant geometric mean regressions for wet and dry weight as a function of percentage cover and the ratios of ash-free/dry biomass for most of the common southern California intertidal seaweeds and many macroinvertebrates. Density data alone are much less likely to show significant mathematical relationships with biomass because of strong spatial and temporal variation in population size structures. Therefore, measurements of morphological features correlated with body size also should be taken and used together with density data to estimate biomass. Good mathematical relationships should exist between size-related morphometric parameters and biomass for barnacles, mussels, limpets, turban snails, anemones, sea stars, and urchins.

HARVESTING PROCEDURES

Because biomass data are obtained from plots, the problems discussed in chapter 6 concerning parallax and the determination of plot boundaries also apply during harvesting. As described previously, rules regarding the fixing of boundaries and the handling of organisms positioned on plot borders must be established a priori to avoid a potentially important source of sampling error (fig. 7.2). Similar to density determinations, only those sessile organisms and seaweed thalli attached within plot boundaries should be harvested to ensure that quantification reflects the amount of biomass anchored to the prescribed area of substratum. This means that fronds of larger seaweeds that lay over the surface of a plot should not be harvested unless the holdfast is determined to lie within plot boundaries. For mobile invertebrates, determinations of whether to harvest an organism should follow the same rules as those described for obtaining density data. In mussel beds, or at times in layered, foliose algal communities, care must be taken to retain crabs and other mobile invertebrates that can move out of plots during sample collection. When necessary, vertical sides made of metal or wood with a flexible material such as foam neoprene or foam plastic can be affixed to quadrat frames and pressed down onto the substratum to prevent mobile animals from leaving plots during harvesting (Gonor and Kemp 1978). Gonor and Kemp (1978) have suggested using a chemical agent such as ethylene chloride, diethyl ether, alcohol, acetone, or a mixture of either alcohol or acetone and dry ice, in conjunction with such frames to kill or narcotize mobile animals. Investigators also have used hot formalin seawater (Glynn 1965) for this purpose. However, formalin poses obvious safety hazards and killing effects can easily spread beyond the intended area. Other chemicals designed to improve the collection of mobile organisms should be fully assessed for these considerations prior to their use.

Because seaweeds and sessile animals are attached to the substratum, these organisms usually must be removed using a flattened scraping tool such as a diving knife, a "putty" knife, or even a standard kitchen knife (fig. 7.3). The swiftly executed placement of a flexible, thin-bladed butter knife is usually effective for removing limpets, chitons, and other invertebrates that hold fast to the substratum. Organisms should be placed in labeled bags or containers immediately following their removal from plots. If making biomass determinations on most or all species, harvested individuals should be quickly presorted in the field using separate plastic bags for each species or species group (fig. 7.4). Although difficult to use in the subtidal, inexpensive plastic bags are generally more efficient than buckets, cloth bags, nylon nets, or other containers for collecting harvested seaweeds and invertebrates, because bags for individual species can be

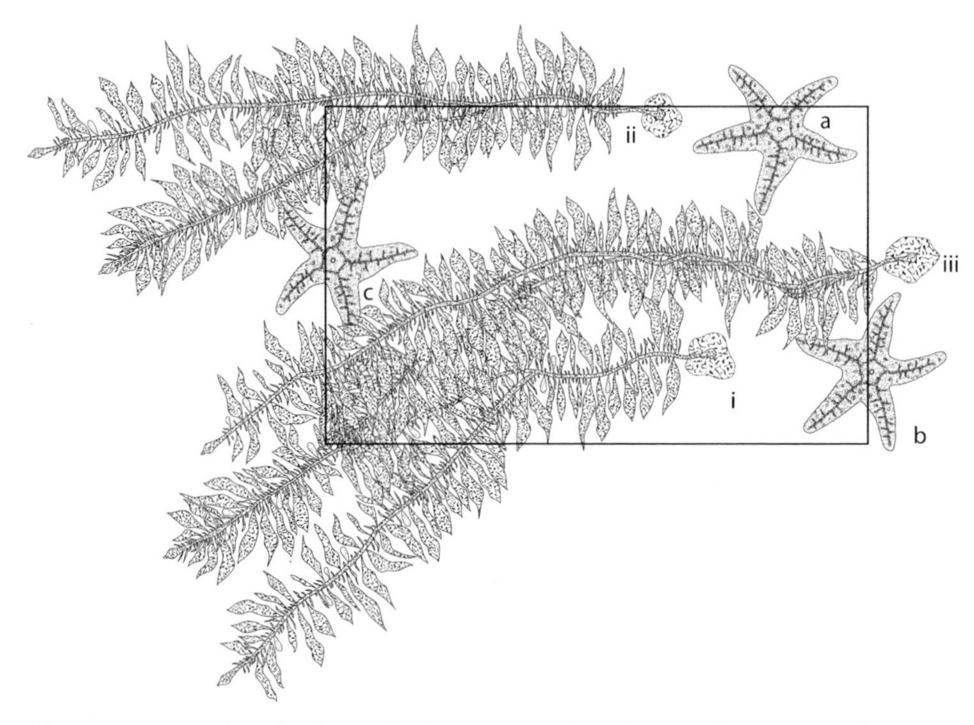

Figure 7.2. Procedure for harvesting large seaweeds and macroinvertebrates from field plots. For sessile animals and seaweeds, only individuals attached within plot boundaries should be removed for most applications. For mobile animals, location at the time of sampling should determine whether an individual is inside or outside of plot boundaries. Rules for including organisms within plots should be established prior to field sampling. For example, if it is decided *a priori* that organisms will be harvested when 50% or more of the individual is determined to occur within plot boundaries, then kelp seaweeds i and ii should be harvested but iii should not. The holdfast for kelp i is located entirely inside the plot, whereas the holdfast for kelp iii is located entirely outside plot boundaries. Kelp ii is harvested because more than 50% of the holdfast is within the plot. In contrast, if harvesting rules include all organisms touching or occurring within the boundaries of two (upper and right) of the four sides of a rectangular or square plot, then sea stars a and b should be harvested but c should not. Sea stars a and b are located on the border of the two sides designated for inclusion, whereas sea star c is located on the border of a nondesignated plot edge.

Figure 7.3. Field harvesting of plot contents. Species are scraped from the substratum using a knife or other flat-bladed tool.

placed in larger bags that contain the contents of specific plots. Also, plastic bags facilitate transportation to the laboratory and storage. However, if specimens are to be analyzed for chemical constituents (e.g., petroleum hydrocarbons or trace metals), or used for other purposes, the choice of harvesting tools and procedures and the composition of storage containers—bags, vials, jars, or other vessels—should be carefully selected to avoid contamination.

A strong disadvantage of destructive sampling programs is that harvested material usually must be stored prior to final laboratory sorting and weighing, which requires some form of sample preservation. For biomass determination, preservation can be accomplished using chemical

Figure 7.4. Field sorting of the contents of harvested plots. Harvested species are bagged separately when possible to facilitate laboratory sorting and reduce laboratory processing time.

preservatives such as formalin (commercial formalin = 37% formaldehyde) buffered with 2 to 3 tablespoons (ca. 0.06 L) of borax per 3.8 L of formalin is an excellent preservative but requires caution: the use of gloves and eye protection when preserving and handling material, fume hoods when sorting or drying samples, and controlled disposal of both liquid and solid wastes. Also, personnel should be educated in how to handle spills. Generally, investigators use a 5%–10% formalin-seawater solution (commercial formalin treated as 100%) for invertebrates (Smith and Carlton 1975) and a 3%–5% solution for seaweeds (Tsuda and Abbott 1985). Seaweeds stay well preserved in formalin-seawater and can be held for long periods without losing color if stored in the dark. Transferring invertebrate specimens to alcohol (70%–95%) is preferred for longer storage, particularly if retaining specimens as vouchers. It might be necessary to first anesthetize delicate, soft-bodied invertebrates or shelled gastropods with tightly closing opercula in magnesium chloride (73 g $MgCl_2 \cdot 6\ H_2O$ per L of tap water) to prevent contraction and ensure complete fixation (Gonor and Kemp 1978).

A viable alternative to the use of chemical preservatives is to freeze algal and invertebrate samples until laboratory processing. Most seaweed thalli and invertebrates respond reasonably well to freezing when the aim is biomass determination, although freezing often negates the use of specimens for detailed taxonomic study or for retention as voucher specimens. Weights of frozen and then thawed specimens, however, will differ from fresh wet weights, particularly for seaweeds or animals whose thalli or bodies begin to show structural breakdowns when defrosted. Because of its simplicity and the ability to avoid using and working with chemical preservatives, freezing is probably the method of choice for most destructive sampling programs if harvested organisms can be returned from the field prior to undergoing spoilage and breakdown. Freezing also generally maintains specimens in a condition suitable for later chemical analyses, while chemical preservation usually compromises their use for these purposes. Hence, if harvested organisms are to be analyzed for chemical constituents, specimens should probably be frozen and stored in laboratory freezers prior to analysis.

BIOMASS DETERMINATION

Wet Biomass

Following the laboratory sorting of harvested material to the desired level of taxonomic discrimination, weights must be determined to obtain biomass data. Ideally, wet weight is the mass of a fully hydrated organism after removing all superficial water adhering to its surfaces or contained inside or between shells, plates, or other structures. Conceptually, wet weights of organisms are simple to determine and require the least processing time. However, because of difficulties in standardizing the degree of hydration and removal of superficial water, wet weights present the greatest potential for measurement error. Consequently, wet weight is usually not recommended for expressing biomass of invertebrates or seaweeds (Gonor and Kemp 1978, Brinkhuis 1985; DeWreede 1985).

For calcareous seaweeds and most invertebrates, wet weights taken of the whole organism will be strongly affected by the weights of calcified walls or tubes, plates, tests, or shells. Moreover, wet weights can yield values of little ecological importance for organisms like anemones, where water retention can vary greatly under different conditions, or species that characteristically have very high water contents and very little usable organic material (Gonor and Kemp 1978). Animals can be removed from their shells, tubes, or plates prior to weight determinations to exclude external inorganic hard parts that make little, if any, energetic contribution to higher trophic levels. Dissection might be required to separate the soft body parts of many species; for some invertebrates, removal from external shells, tubes, or plates can be facilitated by first relaxing animals in a magnesium chloride solution. In some species, placing animals in an acidic dilute formalin solution [(15 mL concentrated glacial acetic acid, 15 mL diluted acetic acid (=commercial vinegar), and 10 mL 10% formalin-seawater)] for several days also can slowly dissolve calcified shells. For calcareous seaweeds, Brinkhuis (1985) suggests dissolving $CaCO_3$ walls of seaweeds by applying HCl (5%–10%, v/v) or by holding specimens in a saturated solution of EDTA (e.g., $Na_2 \cdot$EDTA = disodium salt of ethylenediaminetetraacetic acid), kept fresh by daily replacement, to remove $CaCO_3$ by chelation.

Weight-contributing epibiota and sediments must be removed prior to weight determination and frozen specimens must be treated carefully during thawing to reduce loss of slimy secretions. Preserved samples are usually rinsed with tap water to reduce fumes and remove excess formalin or alcohol before sorting. For some species, significant weight changes can occur during storage in formalin or alcohol, so attempts should be made to process samples within a standardized period to reduce sample-to-sample variation resulting from the effects of preservatives. It is particularly difficult to obtain accurate wet weights of shelled gastropods, which can hold unknown amounts of water inside their shells, and organisms such as barnacles, which are difficult to remove from rocks without losing organic material during collection, storage, or sorting. After ensuring that tissues are fully hydrated and have not desiccated, organisms are blotted to remove surface water and weighed, usually to no more than ± 0.1 g on a top-loading balance. When laboratory sorting takes a long time, samples can be occasionally sprayed with seawater mist or kept moist with wet paper towels to ensure that tissues are hydrated prior to blotting. If necessary, larger organisms can be rehydrated by submersing them directly in seawater for a few minutes. Different techniques have been used to remove the exterior film of surface water from seaweeds, including twirling specimens in mesh bags (Brinkhuis 1985) or in salad spinners. However, repetitive hand-blotting with paper towels until no water appears to be transferred to the paper is probably the simplest technique for removing excess surface water from most specimens.

Dry Biomass

Dry biomass is the weight of tissue remaining after removing all water (Brinkhuis 1985; DeWreede 1985). To obtain dry weights, samples should be cleaned of sediments and epibiota, rinsed quickly in tap water to remove sea salts, placed in preweighed (tared), labeled weighing containers, and put into a drying oven, where they are held at a specified temperature until constant weight is achieved. Depending on the purpose of the study, walls, tubes, plates, tests, or shells made of inorganic material can be removed prior to drying calcareous seaweeds and invertebrates as described previously. Aluminum weighing boats available commercially make good weighing containers. Where samples are too large to fit into standard weighing boats, aluminum foil can be molded around beakers or other objects to make custom-sized weighing containers. Drying can be achieved by heating samples to drive off water or by lyophilization (=freeze-drying). The latter procedure is more time-consuming than oven drying and prevents significant changes in chemical constituents of specimens during the drying process (Brinkhuis 1985). However, for most applications, drying specimens in a drying oven is recommended.

Complete sample dryness is difficult to define, although the goal is to dry the sample to constant weight under the conditions employed. The procedures and temperatures used vary greatly among investigators and must be clearly specified. Brinkhuis (1985) reports the use of drying temperatures ranging from around 20°C to more than 110°C to achieve dry biomass values for seaweeds. Low drying temperatures often are employed when large amounts of material are dried in the open air; however, Brinkhuis (1985) recommends against air drying samples because of difficulties in repeating temperature, humidity, and other conditions that affect dry weight. Gonor and Kemp (1978) have recommended using temperatures of 70°C for drying samples of intertidal organisms, but the most commonly used drying temperature is probably 60°C. This temperature is conveniently achieved in a variety of drying chambers, including handmade containers that rely on heating from light bulbs. Drying temperatures of 100°C to 110°C will accelerate the drying process and are commonly used for obtaining dry weights of terrestrial plants. Although drying at 100°C to 110°C drives off all water (except for bound water and water of crystallization), samples also undergo chemical changes and lose volatile compounds (Brower et al. 1998). Hence, if dried samples are to be analyzed for chemical constituents, drying temperatures should not exceed 60°C (Brinkhuis 1985). Regardless of the method used, samples should be dried long enough to obtain a constant weight, and drying and weighing methods should be fully reported.

Brinkhuis (1985) notes that dry seaweeds, and probably other marine specimens, are very hygroscopic and quickly absorb water (and weight) from surrounding air humidity and even desiccant. Consequently, samples should be transferred to desiccators charged with fresh desiccant immediately after their removal from drying ovens to protect against weight changes resulting from the absorption of water vapor. After cooling to room temperature under desiccant, samples should be weighed to ±0.001 g using a precision balance. A small beaker with fresh desiccant can be stored in the corner of the weighing chamber to obtain stable weights by preventing specimens from absorbing water vapor during the weighing process.

Figure 7.5. Glass weighing jars and ceramic crucibles containing dried algal powder loaded on the shelf of a muffle furnace prior to ashing.

Ash-Free Dry Biomass

Dried organisms also can be combusted to determine ash-free dry biomass or organic dry biomass. Ash-free dry weight is the dry weight of organic material in the absence of inorganic materials and is obtained by subtracting the weight of ash (inorganic material) from the dry weight of the sample (Brinkhuis 1985). Intertidal invertebrates and seaweeds have different body compositions and will retain different concentrations of inorganic salts and minerals in their soft tissues after drying, even after the removal of shells, plates, tubes, or other inorganic structures. Biomass data are perhaps best expressed as organic or ash-free dry biomass per unit area because these data provide a common (organic matter only) basis for describing the standing stocks of all macrophyte and macroinvertebrate populations. However, this means that additional work will be required to determine the contribution of inorganic ash to the dry weight of each sample.

Ash weights of specimens are obtained by combusting known dry weights of samples in a muffle furnace, usually at 500°C to 550°C, for at least 3–4 hr but often as long as 6 hr or more to determine inorganic ash content (Brinkhuis 1985). The specific time needed to oxidize all organic material will vary depending on the temperature and the type and amount of the sample and should be worked out. This is done by reashing and reweighing samples until stable ash weights are obtained. Ashing should be considered incomplete if black charcoal deposits remain in samples (Brower et al. 1998). Oven temperatures should not exceed 550°C to prevent volatilization of sodium and potassium and carbonate combustion (Brinkhuis 1985; Brower et al. 1998).

If shells, plates, tubes, or other inorganic structures are present, these should not be placed in the muffle furnace but should be carefully removed as described previously, scraped to remove organic residues, then dried and weighed separately. Aliquots (0.5 to 2–4 g) of dried seaweed or invertebrate material should then be broken into small pieces, or preferably ground to a powder in a grinding mill, and placed uniformly over the bottom of acid-cleaned and preweighed ceramic crucibles or small, glass weighing jars (fig. 7.5). The dried material or powder should be weighed together with the crucible or jar to at least 0.0001 g using a precision balance noting the weighing precautions described previously. Brinkhuis (1985) notes that crucibles or jars should not be placed directly into a hot furnace but should be added when temperatures are lower than 200°C to avoid rapid ignition and the loss of inorganic material. Combustion time should begin when the furnace has reached the designated temperature. Samples should then be removed from the furnace using tongs, carefully placed in a desiccator to cool, and weighed.

SUMMARY

Clearly, biomass methods of measuring abundance require more laboratory and field time and carry more costs for supplies and materials than nondestructive methods. Biomass data better characterize trophic resource availability than density or cover data and provide a uniform, comparable means of expressing the abundances of both seaweeds and invertebrates. However, biomass determination usually requires that organisms be harvested, resulting in the need to disturb the study site during sampling. It is difficult to envision a case where destructive biomass sampling would be preferred for studies of small, ecologically sensitive habitats or for performance of long-term monitoring programs that require repeated assessments of the same study site. If biomass data are required, costs can be reduced and habitat damage lessened by subsampling species of interest to establish relationships between nondestructive sampling parameters such as cover or size plus density and biomass. Similarly, relationships can be established for each species and used to calculate dry from wet weight or ash from dry weight to save time and money. However, caution should be exercised in using such mathematical functions because of spatial and temporal variations in the established relationships. Because biomass sampling is costly and usually results in habitat disturbance, the investigator should ensure that biomass data are required before initiating a destructive, biomass sampling program. Consequently, destructive sampling for biomass should be carried out only when necessary to achieve the goals of the study and when nondestructive procedures cannot provide the required information.

LITERATURE CITED

Bellamy, D. J., A. Whittick, D. M. John, and D. J. Jones. 1973. A method for the determination of seaweed production based on biomass estimates. In *A guide to the measurement of primary production under some special conditions,* Paris: UNESCO. 27–33.

Brinkhuis, B. H. 1985. Growth patterns and rates. In S. Littler, editors. *Handbook of phycological methods. Ecological field methods: Macroalgae*, ed. M. M. Littler and D. S. Littler, 461–77. Cambridge: Cambridge Univ. Press.

Brower, J. E., J. H. Zar, and C. N. von Ende. 1998. *Field and laboratory methods for general ecology.* 4th ed. Boston: WCB/McGraw–Hill. USA.

DeWreede, R. E. 1985. Destructive (harvest) sampling. In *Handbook of phycological methods. Ecological field methods: Macroalgae,* ed. M. M. Littler and D. S. Littler, 147–60. Cambridge: Cambridge Univ. Press.

Glynn, P. W. 1965. Community composition, structure, and interrelationships in the marine intertidal *Endocladia muricata–Balanus glandula* association in Monterey Bay, California. *Beaufortia* 12:1–198.

Gonor, J. J., and P. F. Kemp. 1978. Procedures for quantitative ecological assessments in intertidal environments. U.S. Environmental Protection Agency Report EPA-600/3-78-087.

Kanter, R. 1978. Mussel communities. Report 1.2. In *Vol. III. Intertidal study of the Southern California Bight.* Washington, DC: Bureau of Land Management, U.S. Department of the Interior.

Littler, M. M. 1979a. Geometric mean regression, product moment correlation coefficient equations for wet and dry weight as a function of per cent cover for biota throughout the Southern California Bight during 1975–78. Vol. IV. Appendix B. In *The distribution, abundance and community structure of rocky intertidal and tidepool biotas in the Southern California Bight. Southern California Bight Baseline Study. Intertidal, Year three*, ed. M. M. Littler, 1–23. Washington, DC: Washington DC: Bureau of Land Management, U.S. Department of the Interior.

———. 1979b. Mean ratio of ash-free to dry biomass for dominant macrophytes and macroinvertebrates with high ash content at sites sampled utilizing the disturbed method in 1977–78. Vol. IV. Appendix C. In *The distribution, abundance and community structure of rocky intertidal and tidepool biotas in the Southern California Bight. Southern California Bight Baseline Study. Intertidal, Year three,* ed. M. M. Littler, 1–5. Washington, DC. Bureau of Land Management, U.S. Department of the Interior.

———. 1980. Southern California rocky intertidal ecosystems: methods, community structure and variability. In. *The shore environment. Vol. 2. Ecosystems,* J. H. Price, D. E. G. Irvine, and W. F. Farnham, 565–608. London: Academic Press.

Littler, M. M., and S. N. Murray. 1974. The primary productivity of marine macrophytes from a rocky intertidal community. *Mar. Biol.* 27:131–35.

Littler, M. M., K. E. Arnold, and S. N. Murray. 1979. Seasonal variations in net photosynthetic performance and cover of rocky intertidal macrophytes. *Aquat. Bot.* 7:35–46.

Poole, R. W. 1974. *An introduction to quantitative ecology*. New York: McGraw–Hill.

Smith, R. I., and J. T. Carlton, eds. 1975. *Light's Manual: Intertidal invertebrates of the central California coast*. 3rd ed. Berkeley: University of California Press.

Tsuda, R. T., and I. A. Abbott. 1985. Collection, handling, preservation, and logistics. In *Handbook of phycological methods. Ecological field methods: Macroalgae*, M. M. Littler and D. S. Littler, 67–86. Cambridge: Cambridge University Press.

Intertidal Black Abalone populations on the outer coast of northwest San Clemente Island, California circa 1972.

CHAPTER 8

Individual-Based Parameters

Age Determination, Growth Rates, Size Structure, and Reproduction

Most rocky intertidal monitoring and impact studies are designed to determine the status of sampled populations solely in terms of abundance. This approach can present two problems. First, abundance data alone do not adequately describe a population in a way that depicts its dynamics. Population density is a result of what Chapman (1985) has referred to as primary and secondary population parameters. Primary population parameters include natality, mortality, immigration, and emigration. Secondary population parameters, such as size and age class distribution and sex ratios, are end products of population dynamics, but also provide a sharper picture of existing structure compared with density data alone. Data on primary and secondary population parameters offer a measure of population status superior to abundance data and provide the information required to compare population growth and structure among sites or over time. Moreover, data on primary and secondary population parameters provide the basis for developing models predicting the future status of populations, including the time course for recovery following an oil spill or other catastrophic event. A second problem involves the high "noise-to-signal" ratio because of the usual high variation obtained when sampling for abundance data or other population-based parameters. In contrast, individual-based parameters, such as size, growth rate, and gonadal production, yield greater statistical power because of lower variability and greater effect size (Osenberg et al. 1996).

The value of primary and secondary population parameters underscores the importance of extending studies beyond the determination of abundance if the goal is to compare the status of populations over spatial or temporal scales. However, studies of primary parameters include assessments of population additions from natality and immigration (in animals) and subtractions from mortality and emigration (in animals).

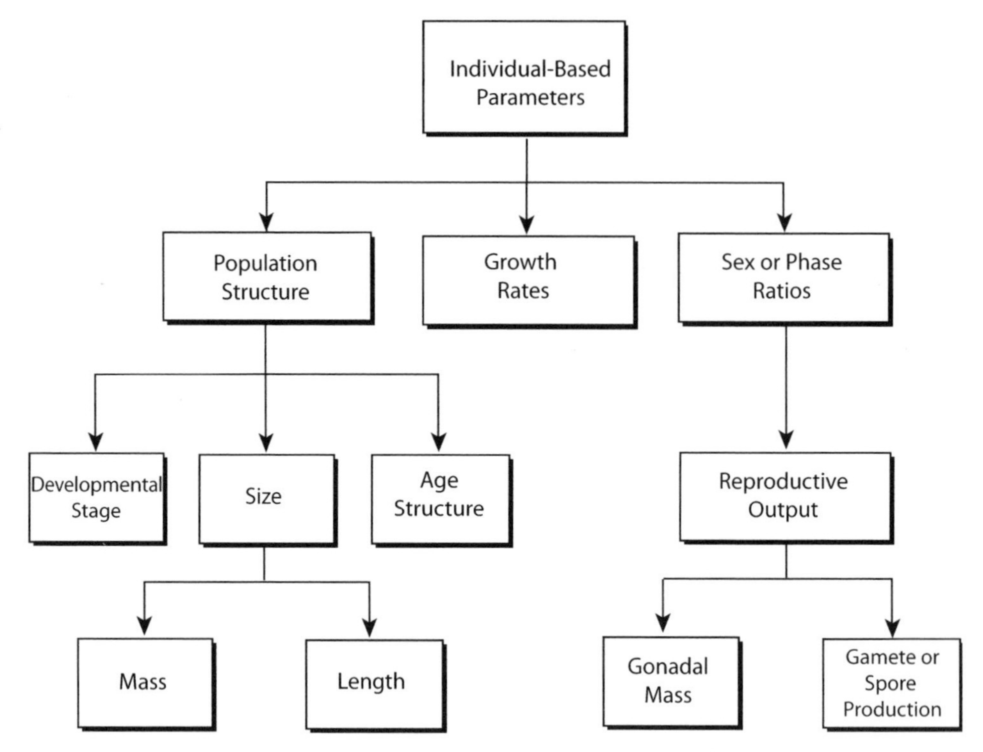

Figure 8.1. Basic approaches for studying secondary population parameters in rocky intertidal macrophytes and macroinvertebrates.

Because most benthic intertidal animals either are sessile or have low mobility, emigration and immigration often are of little importance and can be ignored. Natality and mortality measurements, however, are difficult and costly to make because most rocky intertidal organisms have small planktonic larvae or spores that are transported over scales ranging from meters to tens or hundreds of kilometers. Usually new individuals are first identified as entering a rocky intertidal population when they become visible settlers on benthic substrata and their appearance is usually used as a substitute for natality in demographic studies. Unfortunately, even studies designed to determine recruitment or mortality rates of early benthic recruits are difficult to perform. Moreover, the best means of measuring mortality in most rocky intertidal populations is through the long-term monitoring of marked individuals. Thus, despite their importance, demographic studies dependent on primary population parameters are almost never incorporated into monitoring programs or environmental impact studies of rocky intertidal populations.

Studies of secondary characteristics, such as size or age class distribution and reproductive condition, provide more complete ecological comparisons of population status than those where only abundance data are collected, and usually are simpler to perform than recruitment and mortality studies. Unfortunately, data describing secondary population parameters are seldom collected and little is known about individual-based parameters such as age, size, and reproductive output for most rocky intertidal populations. The purpose of this chapter is to describe and discuss selected approaches for measuring secondary population parameters in rocky intertidal organisms (fig. 8.1). Emphasis has been placed on procedures used to determine age and size class distributions, growth rates, sex or phase ratios, and reproductive condition in seaweeds and benthic invertebrates.

SEAWEEDS

Growth Rates and Age Determination

We know very little about the maximum ages of individual thalli or the age structure of most seaweed populations (Chapman 1985, 1986). In many invertebrates, growth continues over time in a predictable allometric pattern, and when the rate of growth is known, the size of certain anatomical features can be used to estimate age. In seaweeds and surfgrasses, however, overall growth patterns rarely proceed exclusively along single, definable axes, and few species exhibit morphological features that consistently coincide with thallus age. Instead, most seaweeds have highly plastic morphologies shaped by the initiation, growth, and development of multiple modular units consisting of primary and secondary axes. In many species, the irregular production of adventitious and secondary lateral branches also can increase "bushiness" and modify thallus appearance. In addition, seaweed morphologies can vary with exposure to light, wave action, herbivory, or any other parameter that affects axis growth rates and longevity and, therefore, thallus size and shape.

Growth Rates. Seaweed growth rates can be determined in the field or in the laboratory, where plants can be held in culture. The various methods for measuring seaweed growth rates have been thoroughly reviewed by Brinkhuis (1985) and are only discussed briefly here. The most commonly employed approaches involve measurements of changes in seaweed length, weight, or area. Procedures for making wet weight determinations of seaweeds are discussed in chapter 7. As pointed out by Brinkhuis, weighing individual seaweeds in the field presents problems unless thalli can be detached, weighed, and reattached such as

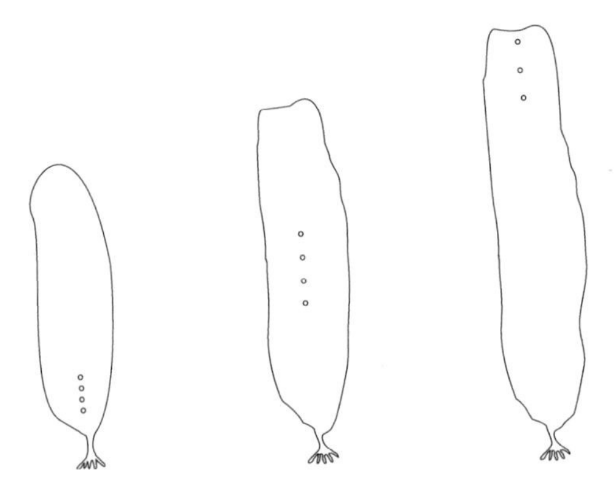

Figure 8.2. Measuring growth in *Laminaria saccharina* by punching a series of small holes (<5 mm in diameter) at fixed distances of 10 cm along the blade. *L. saccharina* blades at time of initial hole punching ($t = 0$) and later ($t = 1$). At each measurement interval, the number of holes remaining, the blade length, the stipe length, and the distance from the base of the blade to the basipetal hole are recorded. (After procedures described by Brinkhuis 1985.)

when plants are clipped to pipes, ropes, or nets and grown in field pens or aquaculture tanks. Consequently, most field measurements of seaweed growth are made from length or area measurements and must be made without detaching plants using nondestructive methods. Generally, this is done by repetitively photographing seaweed axes or whole thalli with a ruler or other item of known dimensions for use as a scale. Measurements are then made on photographic records in the laboratory with rulers, calipers, planimeters, or image-analysis software (see chapter 6 for discussions of photography and computer processing of images).

Because growth of most seaweeds is "plastic" and can proceed in multiple planes, the best candidates for *in situ* growth studies are highly differentiated species that grow principally along identifiable axes. Examples include certain kelps, such as *Laminaria* species, where the linear elongation of blades can be measured (e.g., Lüning 1979; Mann and Mann 1981) by following the migration of holes punched above the localized intercalary blade meristem (fig. 8.2).

Age Determination. The life spans of seaweeds and surfgrasses vary greatly among species and habitats. Some seaweeds, including opportunistic green algae and the summer brown annual *Leathesia difformis* (Chapman and Goudey 1983), are known to have life spans measured in weeks or months. In other species, for example, the kelps *Laminaria saccharina* (Parke 1948) and *Pelagophycus porra* (Coyer and Zaugg-Haglund 1982), very few individuals survive for more than 2 years. Life spans of 2.5 to more than 8 years have been described for Atlantic (*Pelvetia canaliculata* and *Fucus* spp.) and Pacific (*Silvetia compressa* as *Pelvetia fastigiata*) rockweeds (Knight and Parke 1950; Subrahmanyan 1960, 1961; Niemeck and Mathieson 1976; Gunnill 1980). Another relatively long-lived seaweed is the giant kelp *Macrocystis pyrifera*, where the life expectancy of established plants in a southern Californian population has been reported (Rosenthal et al. 1974) to exceed 7 years. Some species of seaweeds may have even longer life spans. For example, David (1943) estimated a 15-year half-life for the Atlantic fucoid *Ascophyllum nodosum*. Moreover, in certain species of red algae with well-developed heterotrichous basal systems, individual genets may persist for decades, or maybe even indefinitely, by periodically producing new erect axes from basal holdfasts. For instance, Dixon (1965) concluded that a *Pterocladiella capillacea* (as *Pterocladia capillacea*) clone was at least 40 years old and hypothesized that one clone of *Pterosiphonia complanata* might be 130 years old. In southern California, the ages of articulated coralline algal genets that produce thick, intertidal turfs from well-developed heterotrichous, crustose basal systems are probably very old, easily exceeding tens of years (Murray, pers. observ.). Crustose algae also may have long life spans under certain conditions. A case in point is the crustose red algal tetrasporophyte phase of *Mastocarpus* (*Petrocelis middendorfii*), where the age of an individual crust has been estimated (Paine et al. 1979) to be as great as 100 years.

For most macrophytes, it is impossible to directly determine age without tagging and following the survivorship of marked individuals. This requires the use of nondestructive marking procedures, followed by the maintenance of marks or tags over the life span of each marked individual. Unfortunately, direct tagging is impossible for most seaweeds because of their soft, fleshy thalli (Chapman 1985), and the direct application of tags usually is avoided because of the high probability of losing both tags and injuring or losing thallus parts during the course of study. Hence, direct tagging mostly has been restricted to larger, tougher species such as kelps (Foster et al. 1985). However, even for kelps, where it is possible to nondestructively tag the thick and tough stipes and holdfast haptera, Foster et al. recommended using double tags—one on the seaweed and one affixed to the substratum immediately adjacent to the holdfast—to increase the likelihood of tag retention. In place of tags, larger, nonturf-forming macrophytes can be relocated by fixing epoxy, bolts, nails, or other reference marks in the substratum adjacent to their holdfasts. In addition, plant relocation can be accomplished by triangulation using transect tapes and two fixed reference points. Differential measurements made with Global Positioning Satellite (GPS) instrumentation also may be useful for obtaining the approximate positions of scattered, conspicuous thalli in heterogeneous intertidal landscapes.

In a very few cases, it is possible to directly determine age by performing measurements on collected plants. For example, in some kelps, including *Laminaria* species (Kain 1963, 1971; Klinger and DeWreede 1988) and *Pterygophora californica* (DeWreede 1984, 1986), age can be determined with some degree of confidence by counting annual growth lines found in the stipe. This procedure requires destructive sampling, however, because stipes must be sectioned to inspect growth rings. Nondestructive estimates of thallus age can be made for species such as the subtidal red seaweed *Constantinea subulifera*, by counting the stipe scars remaining from annual cycles of blade production (Powell 1986). Age estimates also can be made on field plants of the Atlantic fucoid, *Ascophyllum nodosum*, where a single air bladder forms each year on growing, upright axes following the first year of development (Baardseth 1968). However, this method is not foolproof, because, as pointed out by Cousens (1981), axes of this brown seaweed often are broken and the number of bladders produced can be used only to calculate a minimum age.

Population Size Structure

Instead of calendar age, a more meaningful approach for studying the population structure of plants is to use size or developmental stage as the incremental unit; in this procedure, individual plants are assigned to size or developmental categories based on a combination of quantitative and qualitative parameters (Hutchings 1986). Not only is this approach usually more feasible for terrestrial flowering plants that also are difficult to age, but it has been argued (Werner and Caswell 1977) that population models based on size or developmental stage are more informative than models based on calendar age for at least some species. As pointed out by Harper (1977) for forest trees, however, a consistent relationship between the size of a plant and its age should not be assumed and small-sized individuals can be of any age.

Categorization by size or stage is an attractive alternative for studying population structure in macrophytes because, as discussed, few species have morphological or developmental characteristics that consistently enable age determination. Except for many of the morphologically complex kelps, rarely do seaweeds and surfgrasses show growth patterns that enable their assignment even to specific developmental stages. This means that size must be used to categorize individuals when performing analyses of population structure. Although demographic events are probably better described by size than by age in seaweeds, this has yet to be unequivocally demonstrated (Ang and DeWreede 1990).

The morphologies and growth characteristics of many macrophytes can confound the development of size-structure classifications and may dictate the use of several morphological parameters to obtain a meaningful system of size categorization. Models for following size classes over time also can be difficult to employ, because unlike unitary invertebrates, seaweeds and surfgrasses can readily decrease in size due to axis loss or breakage. Therefore, between measurement periods, seaweeds can move both backward and forward through size-based categories unlike age-based categorization. For these reasons, the population structures of seaweeds and surfgrasses are much more difficult to study than those of invertebrates, and information on size, age distribution, or other secondary population parameters exists for only a few species.

Axis length is the most common thallus parameter used to establish size classes for demographic analysis of seaweed populations. The lengths of individual axes can be determined directly in the field for many species without destructive removal from the substratum. The best candidates for studies of population size structure are fucoids, kelps, and other large, conspicuous macrophytes, where distinct individuals can be recognized and major axes easily identified and measured. For example, Ang et al. (1996) used maximum axis length to group plants by size class in their analysis of variations in the structure of populations of the economically important Atlantic seaweed *Ascophyllum nodosum* at sites with different harvesting histories. Similarly, Denis and Murray (unpublished data) used maximum axis length to describe changes in the structure of *Silvetia compressa* (=*Pelvetia compressa*) populations following the application of experimental trampling treatments (fig. 8.3). Like most seaweeds and surfgrasses, however, thalli of *Ascophyllum* and *Silvetia* have numerous axes of varying lengths. For these reasons, the length of the longest axis is usually recorded and used to represent thallus size and to place individual macrophytes into size classes.

Unfortunately, maximum axis length may not always be a good correlate of size, particularly if size is being used to represent thallus biomass. This is particularly true for species with plastic morphologies, where the relative lengths and numbers of major and minor axes are known to vary considerably among populations or over time. Populations of these kinds of seaweeds growing in different sites or even different habitats within a single site, will generally show much morphological variation and exhibit different thallus length-to-biomass ratios. Hence, for most purposes it is advisable to work out and report relationships between maximum axis length and biomass when comparing the

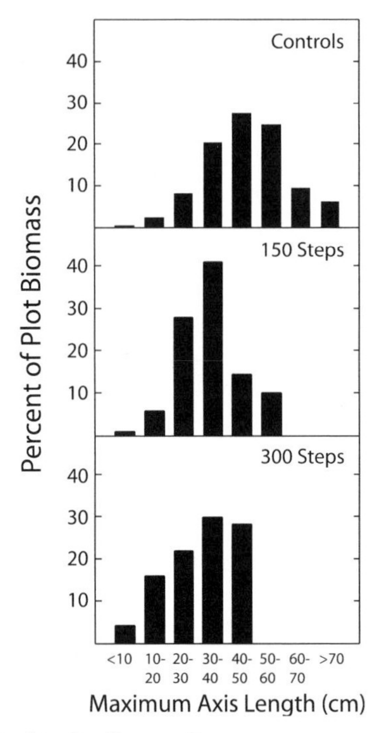

Figure 8.3. Effects of two levels of trampling treatment on population size structure in *Silvetia compressa* at Monarch Bay, California. Depicted is the proportion of the biomass obtained for each size class interval for all thalli harvested from control plots (0 steps) and plots receiving monthly 150- and 300-step trampling treatments. Size classes based on maximum axis length. Thalli were pooled from 0.53 × 0.75-m replicate plots (n = 5) for controls and for both trampling treatments.

structure of seaweed populations, if the analysis is to represent the structure of the standing stock. However, a disadvantage of obtaining length-to-biomass data is that biomass determinations are almost always difficult to perform in the field (Brinkhuis 1985) and require destructive harvesting (see chapter 7).

Sex or Phase Ratios

Most of our knowledge of spatial or temporal variation in patterns of macrophyte reproduction has been gained from largely qualitative phenological studies where the presence or absence of reproductive organs is noted during field surveys or from laboratory inspection of herbarium specimens. Except for obviously fertile specimens, it is usually very difficult to determine even the sex or phase of most seaweeds without the benefit of detailed microscopic examination. This is because seaweed reproductive structures are small and cryptic, and are often found beneath the surface, where they are concealed inside thallus tissues. Although surfgrasses produce flowers, frequently these are absent in local populations during most of the year and can be difficult to observe even when present. Unlike surfgrasses, which produce dioecious flowers, the gametangial phases of seaweeds may be either monoecious or dioecious depending on the species. However, even when male and female thalli occur separately they are almost never sexually dimorphic except for species of the red algal order Palmariales, where dimorphic gametangial thalli characterize the group.

In a few red algal genera, including most species of *Rhodymenia* and *Porphyra*, fertile male thalli can be distinguished readily in the field because visible white or light-colored male sori develop on thallus surfaces. In contrast, male thalli may not be detected easily in those species where male gametangia are scattered over the thallus or develop only within narrow time windows. Female thalli also are difficult to detect in most red seaweeds except following fertilization because female gametangia cannot be seen without microscopic examination (and often laborious specimen preparation). However, following fertilization, carposporophytes develop from zygotes as small, dense filamentous aggregations that often are visible to the unaided eye. In many of these species, the carposporophytes become surrounded by gametangial tissues and form visible, berrylike structures called cystocarps, embedded below or projecting from the thallus surface, which form dark spots ranging in size from 1 to 5 mm in diameter (Santelices 1990). In contrast, gametangia usually are not highly differentiated in green seaweeds, and although mating strains or anisogametes may be present, "true" male and female gametes and phases usually are not designated. This is also the case in many brown algae, except gametangia usually can be differentiated from vegetative cells. Male and female gametangia, however, do occur in certain orders, including the Desmarestiales, Laminariales (kelps), Fucales, and Dictyotales.

Unfortunately, many seaweeds are usually sterile when collected or encountered in the field and show no apparent signs of either gametangia or sporangia. Except for many of the red algae that belong to the Gigartinaceae, phase identification in these cases is almost impossible without laboratory measurements of ploidy level. In many gigartinacean red algae, gametangial and tetrasporangial thalli can be distinguished in the field, even when thalli are sterile. This is done by testing for differences in wall chemistry that occur between gametangial and tetrasporangial phases by adding resorcinol to test tubes containing small thallus pieces and inspecting resultant color changes (Garbary and DeWreede 1988).

Phase determination is most easily made in seaweeds with heteromorphic life histories, such as the kelps, where gametangial and sporangial thalli differ greatly in appearance. In contrast, reproductive structures almost always must be found to make phase determinations in seaweeds with isomorphic gametangial and sporangial phases without resorting to arduous laboratory determination of chromosome number or measurements of relative nuclear fluorescence (see Murray and Dixon 1992).

In the red algae, sporangial plants can be distinguished from gametangial thalli when tetrasporangial structures or sori are present, but rarely without microscopic inspection. In the Laminariales (kelps) and Desmarestiales, the life history consists of heteromorphic phases, and the male and female gametangial thalli are composed of only a few cells and cannot be detected without great effort in the field. This means that all of the large, conspicuous thalli of the kelps or members of the Desmarestiales are sporangial thalli or sporophytes. In contrast, the gametangial and sporangial phases in species belonging to the Dictyotales are isomorphic and the phase or sex of unfertile specimens cannot readily be determined. When fertile, however, male, female, and sporangial thalli can be discriminated with the unaided eye or with the aid of a hand lens based on subtle differences in the appearance of their reproductive organs. Male gametangia occur inside light-colored sori on the thallus surface, whereas oogonial sori are brown and sporangial sori have a brown-and-white speckled appearance due to a mixture of brown sporangia and numerous white sterile hairs. The life history of members of the Fucales does not include true gametangial and sporangial phases but instead consists of only a diploid phase in which male and female gametangia are produced inside pitlike structures called conceptacles. In monoecious species, the sex of a specimen usually must be determined by microscopic examination, although eggs or packets of male gametes may be observed adjacent to conceptacles immediately following their discharge.

Reproductive Output

When fertile, seaweeds can produce very large numbers of reproductive propagules. For example, *Laminaria digitata* sporophytes inhabiting 1 m^2 of rocky habitat can produce more than 20×10^9 spores (Chapman 1984), and enough spores can be released from plants in a *Laminaria hyperborea* forest to accumulate 3.3×10^6 spores mm^{-2} of benthic surface (Kain 1975). Because of the high production of reproductive propagules and difficulties in identifying fertile seaweed tissues, quantitative determination of reproductive production (e.g., spore or gamete output) or the amount of resources allocated to reproduction can be very difficult in seaweeds (Santelices 1990). Among the best candidates for measuring reproductive effort are brown algae belonging to the Fucales. These seaweeds, (e.g., *Fucus*, *Silvetia*, *Hesperophycus*, *Halidrys*, and *Cystoseira*) have simple life histories where the only morphological phase is the diploid, gamete-producing plant. Moreover, the gametangia are confined to conceptacles or pits that occur only on conspicuous, swollen branch tips called receptacles (fig. 8.4). Thus, it is relatively easy to note receptacle

Figure 8.4. Swollen terminal receptacles of *Fucus gardneri*.

presence on plants in the field and to obtain harvested specimens to determine the ratio of receptacle-to-vegetative biomass.

As pointed out by Chapman (1985), the reproductive effort of a seaweed can be determined by measuring the proportion of the total thallus biomass allocated to reproductive functions or to gamete production alone. Unfortunately, few such studies have been performed, and these have produced variable results (Santelices 1990). In kelps, where reproductive structures are concentrated in fertile sori, estimates of resources allocated to reproductive tissues have varied from about 4% in *Macrocystis pyrifera* to 2%–30% in *Laminaria* and 10%–20% in *Ecklonia radiata* (Santelices 1990). Higher estimates have been made for the fucoid *Ascophyllum nodosum* (40%–60%), the crustose red alga *Lithophyllum incrustans* (10%–55%), and the green seaweed *Ulva lactuca* (20%–60%) (Santelices 1990).

In estimating the amount of resources allocated to reproduction in fucoids, receptacle tissues usually are assigned a reproductive function. Using this approach, the percentage of reproductive biomass has been found (Cousens 1981, 1986; Robertson 1987; Back et al. 1991; Ang 1992; Mathieson and Guo 1992; Brenchley et al. 1996) to vary seasonally and among fucoid species. In addition, Russell (1979) suggested that differences in ratios of receptacle-to-vegetative mass in *Fucus vesiculosus* populations from sheltered and exposed shores might result from adaptations to habitat conditions.

Estimates of reproductive effort can be made by calculating the ratio of receptacle to total biomass. For example, Brenchley et al. (1996) determined that mature receptacles accounted for more than 40% of the dry weight of a *Fucus serratus* population during the reproductive season but made up less than 10% of the dry biomass during nonreproductive winter periods. In contrast, Brenchley et al. reported that mature receptacles amounted to more than 90% of the dry biomass in a population of *Himanthalia elongata*, a species with only a single reproductive event. If required, gamete weights also can be estimated using approaches described by Vernet and Harper (1980) and Chapman (1985). Interestingly, estimates of the proportion of total body weight allocated to gametes ranged from 0.1% to 0.4% in three species of *Fucus*, values two orders of magnitude below calculations of reproductive allocation based on receptacle mass (Vernet and Harper 1980).

Receptacles also can be used to estimate the seasonal reproductive effort or to obtain estimates of the number of eggs produced in a fucoid population (see Vernet and Harper 1980). This can be accomplished by first determining in the field the mean number of receptacles or the receptacle mass per unit area of substratum and then estimating the conceptacle density per individual receptacle or per receptacle mass. Mean conceptacle densities can be determined by counting the number of conceptacles on randomly selected receptacles using transmitted light and a dissecting microscope. The number of eggs per conceptacle can be determined by dissecting out a number of randomly selected conceptacles, carefully squashing the conceptacles on a microscope slide, and counting the number of mature eggs using a compound microscope. Using the described procedures, Aberg and Pavia (1997) calculated egg production for eastern North Atlantic populations of *Ascophyllum nodosum* to be as high as 2.5×10^9 eggs m^{-2} during the reproductive season. Using similar approaches, Koehnke and Murray (unpublished data) estimated seasonal variations in receptacle and conceptacle production for a southern California population of *Silvetia compressa* (fig. 8.5).

MACROINVERTEBRATES

Growth Rates

The growth rates of intertidal invertebrates can vary considerably among sites and over time due to changes in food availability or environmental conditions such as temperature. Growth rate studies can be used to estimate the age of intertidal invertebrates by fitting growth rate data to a mathematical function. This is probably the most common method for estimating the ages of most rocky intertidal invertebrates because most species lack annuli or other time-specific shell or anatomical markings that can be used directly for age estimation.

Because macroinvertebrates of different sizes grow at different rates, animals of similar size should be selected for comparative growth rate studies. Smaller animals grow at faster rates and should yield measurable changes in biomass, shell size, or dimensions of selected anatomical features in a shorter period of time. On the other hand, if the study goal is to age animals based on growth rates, then animals ranging widely in size are selected for study in order to obtain the size-related differences required for use in equations describing size-age relationships. In either case, growth rate studies require that the investigator follow marked animals to record changes in measured parameters over time.

Marking Animals with Tags. Because growth rates of animals are determined by making repetitive (at least two) measurements on the same individual, marks must allow for the identification of different animals in the field and remain in place for the length of the study. Keeping marks on invertebrates for the period (usually >6 months) needed to obtain measurable growth increments is often difficult, particularly in rocky intertidal habitats strongly influenced by wave action or sand scour. To prevent tag loss, double and even triple marking is recommended, and frequent tag maintenance and remarking may be necessary.

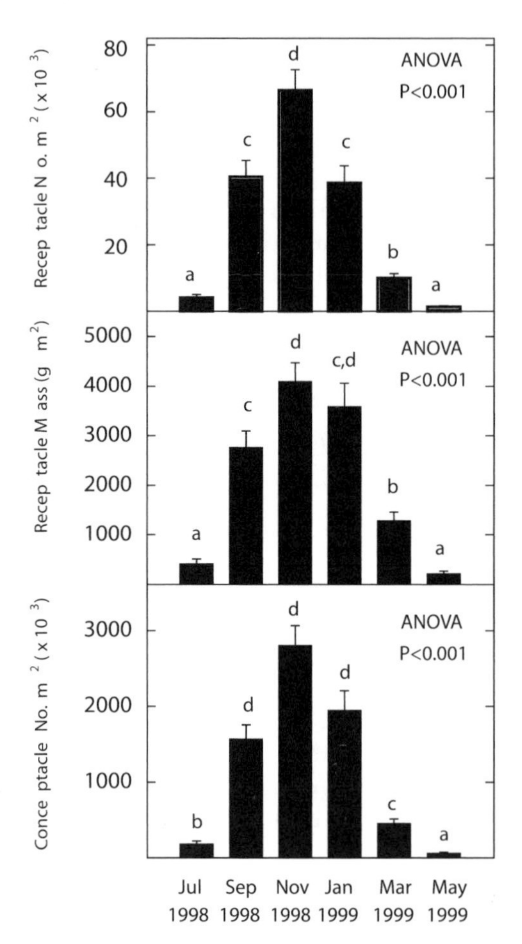

Figure 8.5. Temporal variation in receptacle and conceptacle production in a *Silvetia compressa* population from Crystal Cove State Park in southern California. Top: Number of receptacles per square meter. Middle: Receptacle wet biomass per square meter. Bottom: Number of conceptacles per square meter. Data are means (+1 SE) for 10 replicate 0.25 × 0.30-m plots in habitat containing a minimum of 95% *S. compressa* canopy cover. Conceptacle estimates extrapolated from receptacle numbers and determined from counts of a minimum of 30 receptacles per sampling period.

For sessile animals, the relocation of marked animals can be facilitated by triangulation from reference screws or bolts put into the substratum, and previously marked individuals often can be identified by their positions even following tag loss. In contrast, the movements of mobile invertebrates will be much more difficult to track, and the recovery of marked individuals may be low. Hence, consideration should be given to affixing multiple marks to tagged animals and to marking extra animals in order to accommodate projected losses of marked individuals during the study.

Much attention has been given to the development of tagging methods for fish and various approaches have been compiled and discussed by Nielson (1992). Benthic invertebrates belong to many phyla, however, and, because of different morphological features, require different tagging strategies. Mollusks and barnacles with calcareous hard parts (e.g., shells or plates) usually can be marked without injuring the animal. Numbered tags can be inserted through a thick portion of the carapace to mark large crustaceans such as crabs and lobsters. The chitinous exoskeletons of smaller juvenile crustaceans, such as shore crabs, can be marked with paint. However, as pointed out by Ju et al. (1999), because marks on the exoskeleton will be lost during molting, crustaceans are particularly difficult to tag. Echinoderms, such as urchins and sea stars, and most soft-bodied animals, such as anemones, are much more difficult to mark because tags usually cannot be securely affixed to their external surfaces without causing injury (but see Joule [1983] for marking soft-bodied polychaetous annelids).

Marking benthic invertebrates with calcareous shells or plates usually has been accomplished in two ways: (1) using colored dot patterns made with enamel paint, nail polish, or epoxy, and (2) etching or affixing numbered tags. Individual mollusks can be distinguished using color-coded dots painted on their shells, a commonly used technique that has been employed to mark even very small (<1-mm) juvenile gastropods (Gosselin 1993). For example, under a dissecting microscope Gosselin used six different paint colors to mark *Nucella emarginata* shells with three dot color combinations, a procedure that can generate up to 186 different color codes if two consecutive dots are never the same color. Numbered tags also can be fixed to shells using either Superglue containing cyanoacrylate ester or epoxy cement. Tags, however, should either be coated with clear acrylic or be of a material that can withstand repetitive emersion, submersion, and sand scour to prevent code loss. Tags used for wires (e.g., manufactured by W. H. Brady Co., Milwaukee, WI; sock WM 67-69, Tape B-500+), to label bivalves (Hallprint, Holden Hill, South Australia), or to label insects such as bees (Chr. Graze; 71384 Weinstadt, Germany) all have been used to successfully mark benthic marine invertebrates (fig. 8.6).

Recently, decimal coded wire tags (DCWTs), with individual electronic codes that can be recorded with a scanning instrument (Northwest Marine Technology, Inc., Shaw Island, WA; www.nmt-inc.com/products/cwt/cwt.html), have been used to tag fish and invertebrates.

Figure 8.6. Limpets (*Lottia gigantea*) affixed with multiple tags.

For mobile animals where tagging cannot be done on soft tissue surfaces, an option is to place the DCWTs in the body cavity, if size and anatomy allow. DWCT tags also can be used to mark shelled invertebrates by gluing the tags to the shell, where they are visible in the field. Because of their size and construction, DCWTs can be used with little difficulty, have a very large numerical capacity, and are inexpensive. A disadvantage of using DCWTs is that scanning equipment can be costly (although portable scanning wands are available), and in many applications, tags must be excised, if implanted, prior to being read. The use of DCWTs has worked well for marking crabs (Van Montfrans et al. 1986), lobsters (Krouse and Nutting 1990; Uglem and Grimson 1995), and sea stars (Robles, pers. comm.), and they have been used successfully for many applications with fish. However, DCWT marked organisms can be difficult or impossible to relocate if animals are not also marked with externally visible tags. Without a second conspicuous tag, it may be difficult to determine tag loss, and much field time will be spent testing animals encountered in the habitat for the presence of DCWTs, particularly if only a small proportion of the individuals in a population has been marked.

Measuring Growth. Invertebrate growth rates can be determined nondestructively by repeatedly measuring biomass or increments in the length, width, or maximal diameter of selected anatomical features. If the goal is to compare growth rates at different sites or times, then animals of comparable size should be studied because smaller animals generally grow at faster rates than larger ones. Representative animals within the required size range should be selected at random from the population being studied and either marked in place or carefully removed, tagged, and then returned to their habitats. Growth rates can be determined by first establishing a reference mark on an animal's growing surface using dye or a mark etched into a hardened growing structure such as a shell. For example, in shelled invertebrates, the distance from reference marks etched into the shell at a specific point of growth can be used to measure shell growth rates. This is commonly done for mollusks by etching a fine crosshair into the shell surface along the axis of maximum growth near the growing edge of the shell. For trochids and other coiled snails, this would be the outer or abaxial rim of the shell aperture (but see Vermeij [1980] for a discussion of the complexity of shell growth in coiled gastropods). In limpets, where the growth axis is nearly linear (Vermeij 1980), the mark is probably best made along the midplane near the posterior edge of the shell. Measurements of the distance between the horizontal line and the shell edge are then made using the vertical portion of the crosshair to reestablish the measuring angle during assessments. Shells can be etched in the field with triangular files, diamond-pointed scribes, strong scalpels, or fine, handheld, battery-operated inscribing tools. Because chitons are flexible and adopt body positions that fit the contours of the substratum, an increase in body length can be difficult to determine without detaching animals. Thus, measurements of chiton growth are probably best made using the fourth or fifth intermediate valve, perhaps from the valve edge to the point where the jugum intersects with the edge of the abutting posterior valve (D. Eernisse, pers. comm.).

Animals that lay down calcium carbonate shells, skeletal plates, tests, or spicules also can be marked by exposure to substances such as tetracycline or calcein that produce a fluorescent mark that can be viewed under blue light. For example, tetracycline has been used to successfully mark oysters (Nakahara 1961) urchin tests (Ebert 1988; Ebert and Russell 1993), and abalone (Pirker and Scheil 1993). Calcein (2,4-bis-[*N*,*N*'-di(carbomethyl)-aminomethyl] fluorescein; Sigma No. C 0875 [Moran 2000]) is believed to be superior to tetracycline, however, because it is more readily absorbed and incorporated into calcified structures, shows brighter fluorescence, and is less toxic (Rowley and Mackinnon 1995). Marking calcified structures with a fluorescent dye lays down a reference mark that can be used later to calculate growth increments but does not allow the distinction of individual animals. Calcein marks can be observed even months after removing calcified shells and structures from sacrificed animals.

As described by Moran (2000), a calcein stock solution is prepared by dissolving 6.25 g of calcein in 1 L of distilled water and then buffered to pH 6 with sodium bicarbonate to enhance solubility. The concentrated calcein solution is then added to filtered seawater to achieve a marking solution of 100 mg calcein L^{-1}. Animals are submersed in the calcein seawater solution or injected with sterile calcein solution (ca. 10 mL/kg of body weight). Rowley and Mackinnon (1995) report very successful results from injection, finding that marks appear within a day. For most intertidal species, exposure to calcein is probably best done by submersing animals in a well-aerated, calcein-seawater solution for at least 24 hr. However, exposure trials should be made to determine the treatment times required to produce detectable calcein marks in the animals being studied. Among other invertebrates, calcein has been used as a marker for measuring growth in mussels (Lutz and Rhoades 1980), urchins (Ebert, cited in Rowley and Mackinnon 1995), juvenile snails (Moran 2000), and brachiopods, cockles, and bryozoans (Rowley and Mackinnon 1995).

Marking calcified shells, tests, or other structures with calcein or other fluorescent dyes has certain disadvantages that need to be considered when performing growth studies. First, the fluorescent mark laid down at the time of calcein or tetracycline exposure can only be viewed after cleaning and often cutting open the shell, test, or other calcified structure, a process that requires sacrificing the marked animal. To view calcein marks, soft tissues are removed (e.g., with sodium hypochlorite bleach) and the calcified structure is placed under blue light (470–510 nm; optimal, 497 nm) so that the incorporated calcein fluoresces yellow–green (514 nm) (Rowley and Mackinnon 1995). If the structures are small or the marks inconspicuous, observations can be made with a fluorescence dissecting microscope, with a standard dissecting microscope using epi-illumination and filters (see Moran 2000), or under high-intensity blue or ultraviolet light using an interference filter (514 nm). Because the fluorescent imprint is not visible in the field, a second tagging procedure should be used to avoid collecting both treated and untreated animals to find calcein-marked organisms.

A disadvantage of shell-based growth measurements is that the individual animals comprising a population may have different shell shapes and thicknesses that might compromise among-site growth comparisons based only on single shell attributes. Because shell measurements may not always accurately represent body size, animal biomass also has been used to measure size and follow growth. If total mass is measured in shell-bearing species, however, the attained weights will be dominated by shell mass, not by the organic weight of the animal. To more accurately obtain measurements of animal mass and eliminate sources of variation due to differences in shell shape and thickness, Palmer (1982) developed a nondestructive technique to separate body mass from shell mass when working with mollusks. This procedure involves taking the weight of the whole animal (shell and body) when submersed in seawater and subtracting this from the weight of the whole animal (including shell) obtained in air. A correction factor is then determined by establishing a regression of immersed shell weight against destructively determined shell dry weight and used to calculate true shell weight under submersed conditions. In performing this procedure, Palmer notes the importance of removing gas from inside the shells and mantle cavities of animals prior to determining their weights in seawater in order to avoid significant weighing errors. This is accomplished by completely submersing animals in seawater for 24 to 48 hr prior to weight determination and by pressing the tips of forceps or other blunt objects against the operculum in gastropods to push the body back into its shell. Although Palmer's procedure has the advantage of being nondestructive, its use is much more time-consuming than simple shell length, width, or diameter measurements, which are more commonly used in field studies of molluscan growth.

Age Determination

Most rocky intertidal macroinvertebrates are generally impossible to age without following marked individuals. Because of their longevity and difficulties in retaining permanent marks, it is usually impractical or impossible to follow the survivorship of marked individuals from recruitment to death. Hence, ages of intertidal macroinvertebrates are usually determined indirectly in one of two ways: (1) calculating age from counts of growth markings or rings or (2) estimating age from growth rates.

Annuli and Growth Rings. Age can best be estimated in those species that produce shells or other persistent structures with markings that show variations in growth rate over the course of the year. If such growth rings or marks result from seasonal variation, growth increments can be calculated from the spacing between markings and age can be determined from counts of marks or rings. In some temperate bivalves (Seed 1980), visible shell markings or annular rings (annuli) are produced in the winter or at the onset of new spring growth; in certain cold-water species shell markings can even result from inhibition of growth during summer when temperatures are high (Wilbur and Owen 1964). Depending on the species and the structure being examined, rings or markings can be viewed and counted, and inter-ring distances measured either by direct observation or by using transmitted light or x-ray photographs.

Unfortunately, the use of growth annuli or seasonally produced markings is not applicable for determining the ages of most rocky intertidal invertebrates because there is no evidence that annular or seasonal markings are consistently produced. An exception is *Tegula funebralis*, which, along the Oregon and Washington coasts, produces annual shell markings along the body whorl (Frank 1975). These markings are known to represent periods of interrupted growth; however, south of Monterey, annuli cannot be consistently detected in *T. funebralis,* presumably because of the absence of regular seasonal interruptions in the annual growth cycle (Frank 1975). Other types of markings that can be used to age benthic macroinvertebrates include lipofuscin accumulation in blue crabs (*Callinectes sapidus*) and other crustaceans (Ju et al. 1999), ligament scars on the oyster *Crassostrea virginica* (Wilbur and Owen 1964), opercular rings in the gastropods *Babylonia japonica, Lithopoma undosum (=Astraea undosa),* and *Turbo setosus* (Wilbur and Owen 1964; Sire and Bonnet 1984; Cupul-Magaña and Torres-Moye 1996), and lines on the plates of the chiton *Chiton tuberculatus* (Wilbur and Owen 1964).

Age Calculation from Growth Rates. By following marked animals over time, either by measuring increases in length, width and diameter of shells and tests, or changes in biomass, rates of growth can be used to predict age based on animal size. Analysis of growth increment data, however, can be complicated because, as mentioned, animals of different size grow at different rates. Under most conditions, smaller and younger animals grow much faster than larger and older animals. Additionally, growth often differs over the various parts of the body so that relative body proportions (i.e., the allometry of the individual) can change over time. In order to develop mathematical relationships that translate the size of an individual into an estimated age, growth data usually need to be obtained from animals that encompass the full range of sizes in the population. Therefore, at the outset of the study, consideration should be given to selecting animals that represent all size classes in the population. Animals are then marked or tagged, the growth parameter is selected (e.g., shell length, width or diameter) and measured at time t, and then animals returned to their natural habitat. After a sufficient period of time, $(t + 1)$ for example, 6–12 months for many intertidal invertebrates, tagged individuals are recovered and a second measurement is made to obtain the growth increment over time (e.g., weeks or months) or the average growth rate. In most cases, measurements need to be made at more frequent intervals, particularly if the goal of the study is to determine seasonal growth rates. Growth functions can then be used to calculate relationships between size and age. Although several growth functions are available, three are probably the most commonly employed: (1) the von Bertalanffy curve, (2) the logistic equation, and (3) the Gompertz function. Detailed discussions of these growth functions are given by Ebert (1999) and are discussed only briefly here, with emphasis on the von Bertalanffy function, which has been applied to limpets (Balaparameswara Rao 1976; Branch 1981).

Following procedures outlined by Balaparameswara Rao (1976) for the limpet *Cellana radiata*, the growth increment (shell length) data are first plotted as length at time $t + 1$ and length at time t to create a Ford-Walford plot. The regression equation for these data is then calculated:

$$S_{t+1} = m\,S_t + i \tag{8.1}$$

where S_t is the size or shell length at time t; S_{t+1}, the size at time $t + 1$; m, the slope; and i, the intercept.

The parameters in Equation 8.1 are needed to calculate S_∞, the theoretical maximum size attained by members of the population based on the obtained growth data.

$$S_\infty = i/(1 - m) \tag{8.2}$$

Many of the most widely used growth curves, including the von Bertalanffy function, are based on the premise that the growth of individuals in the population is determinate and reaches some maximum size (e.g., S_∞). The slope derived from Equation 8.1 is then used in Equation 8.3 to calculate K, the growth coefficient:

$$K = (-\log_e m) \tag{8.3}$$

K is then used in the von Bertalanffy growth equation (Equation 8.4) to determine the size–age relationship for the population, where t_0 is the theoretical age at which the size is zero:

$$S_t = S_\infty\,[1 - e^{-K(1 - t_0)}] \tag{8.4}$$

Of course, age estimates made from growth equations are only as reliable as the growth function itself, and several assumptions are made by the von Bertalanffy equation besides determinate growth leading to a maximum theoretical body size, including extrapolation of the age at which size is zero. In addition, the theoretical maximum size (S_∞) obtained for the population of interest using a Ford-Walford plot may actually be less than that known to occur for the species. Using the von Bertalanffy equation, Balaparameswara Rao (1976) estimated that the largest *C. radiata* on Indian shores have survived for 3 to 4 years. Similarly,

Kido and Murray (2003) determined that the ages of the largest individuals in an open rock population of the exploited owl limpet (*Lottia gigantea*) exceeded 8 years in southern California. Excellent discussions of growth functions and their applications are given by Ebert (1999).

Population Size Structure

Size-Frequency Profiles. Since the ecological functions performed by intertidal invertebrates probably are best correlated with size, comparisons of size-frequency profiles among sites or over time are more informative than abundance data alone. Size-frequency profiles can be used to gain insight into important population features such as age structure and recruitment patterns, and can provide the basis for estimating survival and growth rates and for building population models (Cerrato 1980; Ebert 1999). Size-frequency profiles also can form the basis for determining recovery time following an oil spill—not just of population numbers but also of population structure and ecological functions tied to structure.

Size frequency profiles are constructed by taking measurements of size-informative parameters for individuals in a population. To be of greatest value, the parameter selected for size measurement will have a strong positive relationship with growth and reflect size-related ecological roles of the organism. Besides biomass, such parameters include, for example, shell length, height, width, and maximum diameter in mollusks, test diameter in urchins, longest arm length or greatest tip-to-tip span in sea stars, and carapace width in crabs. Choosing the length or appropriate parameter for measuring shell size in mollusks must be given careful consideration. For limpets, the maximum shell length along the anterior-posterior axis has routinely been used to represent size. However, the allometry of coiled gastropods (Vermeij 1980) can present problems in determining shell height, which is measured along the axis of coiling. In addition, shell apexes often become eroded, making length measurements inaccurate. Consequently, maximum shell width (=distance at right angles to the shell axis) is often used to indicate shell size in coiled gastropods.

Excellent discussions of techniques for deducing attributes of molluscan populations from analyses of shell parameters are given by Cerrato (1980) and Ebert (1999). To build size-frequency profiles of macroinvertebrate populations, individuals are randomly sampled from the selected population, and accurate measurements made on the anatomical parameter chosen to represent size. Care must be taken to ensure that smaller, less conspicuous members of the population are not overlooked during sampling and that, if juveniles and adults occupy different microhabitats, this is taken into account. Digital calipers can greatly increase the rate at

Figure 8.7. Performing size measurements of owl limpets (*Lottia gigantea*) in the field.

which size measurements of most common intertidal invertebrates can be made in the field and reduce errors, particularly when multiple investigators perform measurements (fig. 8.7). For some invertebrates, such as barnacles, measurements must be made on animals fixed to the substratum, but it is best to leave even animals such as limpets and chitons in place when performing measurements, to avoid injury. For other invertebrates, such as turban snails and littorines, animals can be picked up, measured, and then returned to the location from which they were collected. When the substratum topography or the presence of other organisms prevents the direct use of calipers, measurements can be taken on dimensions obtained with dividers that can more easily be manipulated around animals in the field.

Following measurement, animals are then sorted into size classes and the percentage or proportion of individuals in the sampled population that occur within each size class is determined (fig. 8.8). The number of size classes and the size range within each size class are selected by the investigator so as to best represent the structure of the population. Cerrato (1980) reports that sizes ranging from 1% to 5% of the maximum size found in the population are mostly used to define size class intervals and suggests that this size range generally provides the number of intervals needed to resolve size classes in the population. Ebert (1999) indicates that at least 15 and preferably 20 or more size classes should be used for some analyses of size-structure data.

Several approaches can be used to compare size structures between and among sites or over time. Three such methods include the following:

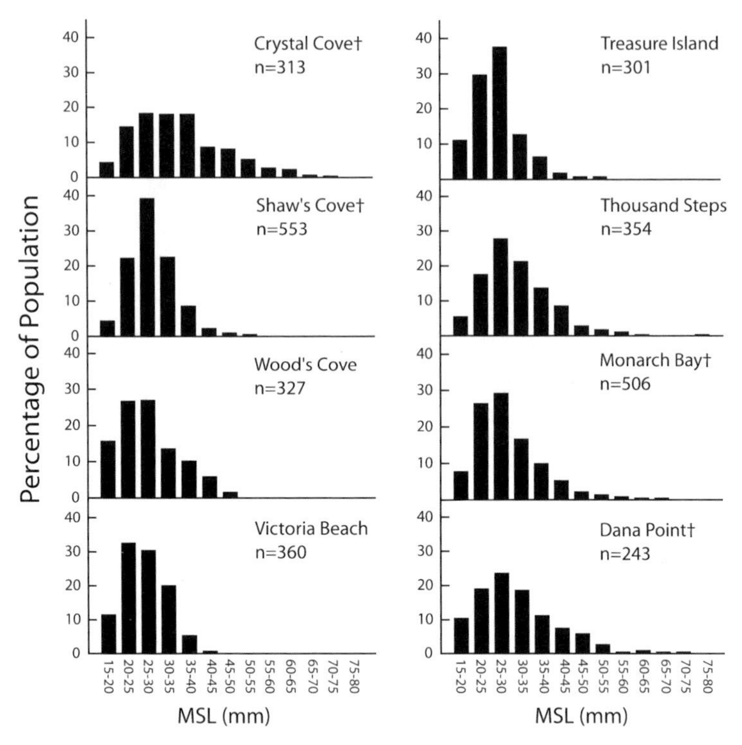

Figure 8.8. Size-frequency plots for the owl limpet *Lottia gigantea* at eight southern California study sites based on maximum shell length (MSL). The indicated number of animals (n) at each site was sorted by 5-mm size classes. Size-frequency distributions differed significantly among sites based on a G-test procedure (see Zar 1999, 505). Largest individuals were found at sites receiving the least human visitation and collection pressure. (Source: Kido and Murray 2003.)

(1) mean sizes can be compared using standard univariate procedures such as analysis of variance or two-sample tests; (2) size distributions can be compared using the Kolmogorov-Smirnov or a similar procedure, or if the data allow, two-parameter gamma curve-fitting can be used to compare profile means and distributions; and (3) the chi-square test can be used to compare the size-frequency distributions. Regardless of the approach, however, caution must be taken when interpreting differences in size-frequency profiles. For example, size frequencies may differ greatly among sites where recruitment and growth rates vary with local conditions. Moreover, the size structure of a population can even vary considerably within a site, such as when different microhabitats are occupied (Kido and Murray 2003) or when mobile invertebrates settle at one

end of their vertical range and migrate to the other. Therefore, it may not be possible to extrapolate population structures measured at one site to another.

Growth Rates from Size-Frequency Profiles. Size-frequency data collected for the same population over a series of time intervals can be used to estimate certain demographic parameters including the growth rates of intertidal invertebrates (Seapy 1966; Ebert 1999). The series of modes obtained from the size-frequency profiles determined for each sampling period are presumed to represent different age classes. Growth rates for different age classes are then estimated by calculating the shift in modal length of an age class as it has progressed from one measurement period to the next. Usually, large sample sizes are required to provide distinct age classes for this procedure. As pointed out by Wilbur and Owen (1964), best results are obtained when populations show seasonal growth and each year's recruitment class is represented by a tightly defined size group. Estimation error increases greatly when recruitment is continuous or occurs over several months or where environmental factors make age classes difficult to distinguish. The absence of a particular recruitment class also can make size frequency–determined parameters difficult to interpret.

Cerrato (1980), Grant et al. (1987), and Ebert (1999) provide excellent discussions of the various approaches for analyzing size-frequency data to make estimates of growth rates and other demographic parameters. Grant et al. (1987) recommend the use of an optimization method such as that described by Macdonald and Pilcher (1979), and describe the Peterson method as the simplest means for separating age classes. In this method, each of the modes in the size-frequency histogram is interpreted as representing a single age class and all age classes are assumed to occur in the sample. Other analytical methods for breaking down size-frequency distributions also are available, but these usually assume that the profiles of individuals in each age class approximate a normal distribution. Graphical methods of analysis (see Cerrato 1980) may not always be reproducible and should not be used except under conditions where distributional modes are distinct and sample sizes are large (Grant et al. 1987).

Reproductive Condition and Output

With the exception of certain invertebrates that are either simultaneous (e.g., barnacles and opisthobranchs) or sequential (e.g., certain patellid limpets) hermaphrodites, most common intertidal invertebrates are characterized by separate sexes and are dioecious or gonochoristic. Interestingly, sexes may not be equally represented in dioecious species, particularly in mollusks, where females tend to be more numerous than males, with this disparity increasing with population age (Fretter and Graham 1964). Because sexual dimorphism is not obvious in most dioecious species, gender is difficult, if not impossible, to determine without killing animals and observing internal sexual structures. Animals are dissected to find male or female sex organs or to observe male and female gonads, which usually differ in shape and color. Because animals usually must be sacrificed, determining the gender composition of a population or establishing relationships between gender and parameters such as size and age is time-consuming and, for most species, is probably best and most efficiently done in conjunction with a destructive sampling program.

In a few kinds of invertebrates, sex can be determined without sacrificing individuals. For example, crabs and many other crustaceans have external features that can be used to indicate sex, and in some dioecious gastropod mollusks, such as abalone, distinctive gonadal material can be observed protruding from the pallial cavity when an animal is very ripe. In some other gastropods, subtle differences between males and females occur in shell characteristics and these differences can be used by trained observers to sex specimens (Fretter and Graham 1964). Occasionally, sex can be differentiated in very ripe sea urchins by sharply shaking animals to discharge eggs and sperm (C. Lambert, pers. comm.).

Gender determination by biopsy can be an attractive and potentially nondestructive alternative for certain dioecious invertebrates, but this procedure requires time-consuming handling of animals. Individuals also will likely be lost to injury from biopsy procedures, with smaller animals probably more vulnerable than larger ones. Also, biopsy sampling is particularly difficult to perform nondestructively in mollusks because they seal punctures to their tissues so poorly. Biopsy sampling can be performed in sea stars and limpets by removing a tiny plug from the gonad and then inspecting the tissue for color or the presence of eggs and sperm. For patellacean limpets, Wright and Lindberg (1979) and Lindberg and Wright (1989) have pioneered a nonfatal technique for gender determination. This involves carefully removing a small amount of gonadal material by inserting a fine syringe needle through the body wall at a prescribed angle. Another approach, used by Frank (1965) to sex turban snails, is to drill a small hole through the shell to reveal the gonad. By following snails whose shells had been drilled and then sealed for 2 years, Frank determined that growth was not significantly affected by this procedure.

Besides sampling intertidal invertebrate populations to determine gender composition or to compare size structures or growth rates of males and females, inspection of gonadal material can be used to determine reproductive (spawning) periodicity and spatial variations in the degree of gonadal production among populations. Seasonal or spatial differences in reproductive condition can be determined by collecting specimens and returning them to the laboratory, where they are dissected and their gonads inspected and weighed. If analyses cannot be done immediately, animals can be frozen or preserved, with the latter approach generally yielding better visual discrimination while also providing the ability to obtain quantitative data on gonadal mass and egg production. Reproductive condition can be categorized by subjectively grading the state of gonadal development (Seapy 1966). Alternatively, the mass or volumetric ratio of gonadal-to-nonreproductive (somatic) body tissue can be determined separately for each sex to provide quantitative information describing seasonal or spatial variation in reproductive condition or age- or size-related patterns of fecundity. If required, the fecundity or the quantity of egg production in the population can be determined by inspecting the ovaries of females microscopically. When using this procedure, eggs can be further categorized in terms of the stage of maturity or development.

Subjective grading of the state of gonadal development is usually based on shape, color, and the quantity of observed gonadal material. These data can be used to determine the frequency of spawning or differences in spawning periodicity among populations. Individuals sampled from the population are dissected to expose gonadal material, which is then graded according to a predetermined qualitative scale. Care must be taken to ensure that the appearance (size, color, shape) of gonadal material is not altered by maintenance (e.g., holding animals in aquariums) or preservation procedures (e.g., fixation, freezing). However, in patellid limpets and archaeogastropods, simple qualitative observations of the shape and color of the ovary alone may not be good indicators of reproductive cycles (Creese 1980).

Quantitative determination of gonadal size can be used to estimate the reproductive output of the population. Working on the assumption that the quantity of gamete production is strongly related to gonadal size, determinations of gonadal mass or volume can be used to estimate fecundity. However, because fertile intertidal invertebrates generally show a strong correlation between body size and gonadal size, gonadal mass or volume should be determined together with the mass of nonreproductive body tissues (fig. 8.9). These determinations can be facilitated in mollusks by preserving and storing specimens in 10% formalin–seawater, which will cause the gonads and other tissues to harden and simplify extraction and weight determination. To ensure best preservation in snails with effective opercula, animals should first be relaxed in magnesium chloride solution so that body tissues are evenly exposed to the fixative. Creese (1980) reports that the wet body weight of limpets

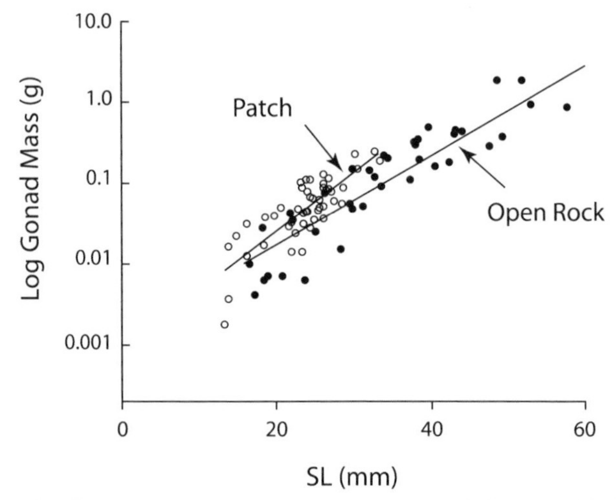

Figure 8.9. Plot showing exponential increase in gonadal mass with shell length (SL) in owl limpets (*Lottia gigantea*). Log gonadal wet mass is plotted against SL for *L. gigantea* obtained from patch (open circles) and open rock (filled circles) habitats at Monarch Bay in southern California. Significant regressions were found for both the open rock ($y = -2.87 + 0.055x$) and the patch ($y = -3.06 + 0.073x$) subpopulations. (Source: Kido and Murray 2003.)

decreases by about 2% following preservation so weight changes due to preservation should be determined. For mollusks and many other invertebrates, the soft body parts of animals must first be removed from their shells or tests. For gastropods, shells can be cracked using a vise or cut open, or animals can be boiled, a procedure that often detaches the retractor muscles holding the body to the shell. Alternatively, the calcium carbonate component of molluscan shells can be slowly dissolved by holding animals for a few days in an acidified formalin solution (15 mL concentrated glacial acetic acid, 15 mL diluted acetic acid [=commercial vinegar], and 10 mL 10% formalin-seawater). The decalcified shells will then consist only of an organic matrix that can easily be cut open to remove the organic contents.

Following dissection and separation, the wet or dry weights of gonads and somatic body tissues should be determined as described in chapter 7. Obtaining dry weights will often produce better precision by reducing weighing errors due to tissue retention of external water or fixative but also will require more time. However, gonads should not be dried if ovaries are to be examined microscopically and eggs counted. Qualitative assessments of reproductive condition can be performed by smearing small amounts of ovarian tissue onto a microscope slide and, upon microscopic examination, grading the sample based on whether it contains mature, immature, or spent oocytes. Alternatively, quantitative determinations of fecundity can be made by microscopically inspecting a known mass of female gonadal material sliced from the ovary and counting and classifying eggs according to their state of development. If the counts obtained for this subsample of ovarian tissue are expressed per milligram of ovarian tissue and are representative of the ovary as a whole, the fecundity of each examined female can be calculated based on its total ovary weight. Together with size-frequency data, this information can be used to calculate the total number of eggs potentially produced by female members of the sampled population. For example, based on size data and quantitative analyses of gonadal tissues, Creese (1980) determined that the average female of the Australian limpets *Notoacmaea petteridi* and *Patelloida alticostata* could produce from 400 to 29,700 eggs and large animals could release 30,000 to 45,000 eggs per year.

A gonadal-somatic index (GSI) can be calculated from gonadal and somatic weights to describe the reproductive condition of animals in the population. This index can be expressed either as the ratio of gonadal mass to somatic body mass or, alternatively (see Creese 1980), as the value of the slope of the linear regression of gonadal mass on somatic mass. Usually, the GSI is expressed as a percentage when calculated as the ratio of gonadal-to-somatic weight. GSI values can be determined separately for each sampled individual and mean values calculated for the population as a whole or, if informative, for each defined size or age class. Mean GSI values calculated for different periods also can be used to determine temporal variation in reproductive development and to document spawning periodicity in the population. If gonads are graded according to oocyte status, gamete output can be calculated by establishing regression equations that describe the relationship between somatic body mass and gonad mass under conditions when the ovaries are both laden with mature oocytes and also after spawning, when only immature oocytes are present. The difference between these two gonadal–somatic mass relationships will provide an estimate of gonadal output. Together with size- or age-structure data, GSIs also can be used to estimate changes in the reproductive output of the population resulting from changes in size or age structure. For example, Branch (1975) determined that gonadal output for the South African limpet *Patella concolor* was reduced by approximately 90% at a site where humans had removed the larger animals from the population for consumption. Hence, GSI data can be used to compare the reproductive conditions of populations at different sites to assess, for example, the impact of an oil spill or the effects of human collecting.

SUMMARY

Abundance data usually provide the foundation for most monitoring and impact studies, yet these data generally have a high "noise-to-signal" ratio and do not fully describe the status (or the ecological services and functions) of rocky intertidal populations. Data describing primary population parameters, such as recruitment and mortality rates, or secondary parameters, such as size and age class distribution, sex ratios, and gonadal production, provide stronger descriptions of population status and dynamics. Moreover, these data may provide more accurate information about population recovery following major impacts such as an oil spill. For example, abundance data might show that cover of an upper shore fucoid or densities of a limpet have returned to the values expected if a spill had not occurred. However, the size or age structures and reproductive outputs of these populations may not yet resemble the structures or outputs of control populations representing their status in the absence of the spill. In this case, species abundances indicate recovery, but the population structures of, and very likely the ecological services and functions provided by, the impacted fucoid and limpet populations have yet to be fully restored.

Unfortunately, data on primary and secondary population parameters are more time-consuming and costly to collect compared with abundance data and, for these reasons, usually are targeted on only one or a few key species and rarely incorporated into monitoring programs. Nevertheless, in monitoring and other long-term sampling programs designed to determine spatial and temporal changes in rocky intertidal ecosystems, efforts should be made to incorporate data collection for at least a few key species on two population parameters: recruitment and size structure. Recruitment of most rocky intertidal species is known to be highly variable in both space and time, and to be closely linked with prevailing oceanographic conditions. A series of poor recruitment years, resulting, for example, from warmer sea temperatures, can lead to changes in population structure and decreases in species abundances even in the absence of major disturbances. Knowledge of spatial and temporal variations in the size structures of targeted populations can indirectly provide information on recruitment and other biological responses to major disturbances or changing environmental conditions.

Valuable information can be collected on population status by taking size measurements of organisms when carrying out rocky intertidal sampling programs because the ecological functions performed by intertidal seaweeds and invertebrates are probably best correlated with their size. Careful consideration must be given to the species selected, how they are sampled, and the parameters (e.g., seaweed axis length or limpet shell length) selected for measurement during program design. Size-structure data can be used to follow temporal patterns in the influx of tiny benthic recruits, to gain insight into population age structure, to estimate its reproductive output, and to otherwise compare the status of populations across sites, over time, or before and after a catastrophic event. Approaches also are available to estimate growth rates and various other population parameters from size-structure data.

The growth rates of rocky intertidal seaweeds and invertebrates vary over both spatial and temporal scales and can lead to significant differences in population structures. Except for a few species (e.g., kelps), growth rates of most seaweeds are difficult to measure in the field because of morphological plasticity and variable growth axes. Growth in most benthic invertebrates, however, is predictably distributed along identifiable axes and can be measured in the field by following marked individuals.

The vast majority of seaweeds and invertebrates are virtually impossible to age using anatomical features. Hence, age must be determined by following marked individuals over time or by estimating age from growth rate data. Marking methods must be carefully considered and customized to the species being studied because tags can cause injury to seaweeds and most soft-bodied intertidal invertebrates and, even in hard-shelled gastropods and bivalves, can easily be lost. Growth rate data can be used to estimate the age of invertebrates using various mathematical functions such as the von Bertalanffy growth equation. However, this approach requires several assumptions and provides poorer estimates for slow-growing older animals that have neared the theoretical asymptotic size for the population. Improved understanding of the growth rates and age structures of most rocky intertidal seaweed and invertebrate populations is needed to determine the effectiveness of natural recruitment cycles and to predict recovery times for returning the ecosystem functions of perturbated species populations following catastrophic events.

Successful reproduction is critical to sustaining a rocky intertidal population over long time scales. Most seaweeds have complex life histories that include both gametophyte and sporophyte phases, and many can perennate or regrow new upright axes from persistent basal systems (=perennation). Seaweeds recruit new individuals into their populations by releasing spores, zygotes, or other propagules that disperse various distances and, also, by regrowing from detached vegetative fragments. Although few quantitative data are available, it is generally assumed that larger-sized seaweeds produce greater numbers of reproductive propagules. Most common rocky intertidal invertebrates inhabiting temperate shores, including most limpets, trochid snails, chitons, littorines, mussels, barnacles, urchins, and sea stars, recruit from planktonic larvae carried variable distances by oceanographic currents. Most also are broadcast spawners that release their gametes into the water column where fertilization occurs. Consequently, fertilization success increases with increasing densities of reproductively mature individuals, and because the amount of gonadal mass increases greatly with body size, larger, more fecund individuals tend to dominate the reproductive outputs of most populations. Hence, populations with large numbers of big individuals produce and export more larvae, and are good sources of larvae for other intertidal sites.

Collecting by humans and other disturbances that kill bigger animals can change the size structure and reduce the reproductive output of an intertidal population. Therefore, data on spatial and temporal variations in sex ratios, production of gonadal mass, and gamete output can be used together with density and size-structure data to evaluate the reproductive status of an intertidal invertebrate population and to determine whether it may serve as an important regional or local source of larvae. Moreover, these data can be used to determine whether the reproductive functions of a population have fully recovered following a catastrophic event such as an oil spill. Unfortunately reproductive data on intertidal populations are time-consuming and costly to obtain, and large gaps exist in our understanding of the reproductive and recruitment cycles and reproductive outputs of most rocky intertidal seaweeds and invertebrates. Although difficult to build in to long-term monitoring programs, studies designed to determine the reproductive status of impacted populations should be considered, where possible, when evaluating the effects of oil spills or other major environmental disturbances.

LITERATURE CITED

Aberg, P., and H. Pavia. 1997. Temporal and multiple scale spatial variation in juvenile and adult abundance of the brown alga *Ascophyllum nodosum*. *Mar. Ecol. Progr. Ser.* 158:111–19.

Ang, P. O. 1992. Cost of reproduction in *Fucus distichus*. *Mar. Ecol. Progr. Ser.* 89:25–35.

Ang, P. O., and R. E. DeWreede. 1990. Matrix models for algal life history stages. *Mar. Ecol. Progr. Ser.* 59:171–81.

Ang, P. O., G. J. Sharp, and R. E. Semple. 1996. Comparison of the structure of the populations of *Ascophyllum nodosum* (Fucales, Phaeophyta) at sites with different harvesting histories. Hydrobiologia 326/327:179–84.

Baardseth, E. 1968. *Synopsis of biological data on* Ascophyllum nodosum (*Linnaeus*) *Le Jolis*. FAO Fisheries Synopsis No. 38. Rome: FAO.

Back, S., J. C. Collins, and G. Russell. 1991. Aspects of the reproductive biology of *Fucus vesiculosus* from the coast of S.W. Finland. *Ophelia* 34:129–41.

Balaparameswara Rao, M. 1976. Studies on the growth of the limpet *Cellana radiata* (Born) (Gastropoda: Prosobranchia). *J. Moll. Stud.* 42:136–44.

Branch, G.M. 1975. Notes on the ecology of *Patella concolor* and *Cellana capensis*, and the effects of human consumption on limpet populations. *Zool. Afr.* 10:75–85.

———. G. M. 1981. The biology of limpets: physical factors, energy flow, and ecological interactions. *Oceanogr. Mar. Biol. Annu. Rev.* 19:235–380.

Brenchley, J.L., J.A. Raven, and A.M. Johnston. 1996. A comparison of reproductive allocation and reproductive effort between semelparous and iteroparous fucoids (Fucales, Phaeophyta). *Hydrobiologia* 326/327:185–90.

Brinkhuis, B.H. 1985. Growth patterns and rates. In *Handbook of phycological methods. Ecological field methods: Macroalgae*, ed. M.M. Littler and D.S. Littler, 461–77. Cambridge: Cambridge Univ. Press.

Cerrato, R.M. 1980. Demographic analysis of bivalve populations. In *Skeletal growth of aquatic animals. Biological records of environmental change*, ed. D.C. Rhoads and R.A. Lutz, 417–65. New York: Plenum Press.

Chapman, A.R.O. 1984. Reproduction, recruitment and mortality in two species of *Laminaria* in southwest Nova Scotia. *J. Exp. Mar. Biol. Ecol.* 78:99–109.

———. 1985. Demography. In *Handbook of phycological methods. Ecological field methods: Macroalgae*, ed. M.M. Littler and D.S. Littler, 251–68. Cambridge: Cambridge University Press.

———. 1986. Population and community ecology of seaweeds. *Adv. Mar. Biol.* 23:1–161.

Chapman, A.R.O., and C.L. Goudey. 1983. Demographic study of the macrothallus of *Leathesia difformis* (Phaeophyta) in Nova Scotia. *Can. J. Bot.* 61:319–23.

Cousens, R. 1981. *The population biology of* Ascophyllum nodosum *(L.) Le Jolis*. Ph.D. dissertation. Halifax, Nova Scotia: Dalhousie University.

———. 1986. Quantitative reproduction and reproductive effort by stands of the brown alga *Ascophyllum nodosum* (L.) Le Jolis in south-eastern Canada. *Estuarine Coastal Shelf Sci.* 22:495–507.

Coyer, J.A., and A.C. Zaugg-Haglund. 1982. A demographic study of the elk kelp *Pelagophycus porra* (Laminariales, Lessoniaceae), with notes on *Pelagophycus* × *Macrocystis* hybrids. Phycologia 21:399–407.

Creese, R.G. 1980. Reproductive cycles and fecundities of four common eastern Australian Archaeogastropod limpets (Mollusca: Gastropoda). *Aust. J. Mar. Freshwater Res.* 31:49–59.

Cupul-Magaña, F.G., and G. Torres-Moye. 1996. Age and growth of *Astraea undosa* Wood (Mollusca: Gastropoda) in Baja California, Mexico. *Bull. Mar. Sci.* 59:490–497.

David, H.M. 1943. Studies in the autecology of *Ascophyllum nodosum* Le Jol. *J. Ecol.* 31:178–98.

DeWreede, R.E. 1984. Growth and age class distribution of *Pterygophora californica* (Phaeophyta). *Mar. Ecol. Progr. Ser.* 19:93–100.

———. 1986. Demographic characteristics of *Pterygophora californica* (Laminariales, Phaeophyta). *Phycologia* 25:11–17.

Dixon, P.S. 1965. Perennation, vegetative propagation and algal life histories, with special reference to *Asparagopsis* and other Rhodophyta. *Bot. Gothoburg.* 3:67–74.

Ebert, T.A. 1988. Calibration of natural growth lines in the ossicles of two sea urchins, *Strongylocentrotus pupuratus* & *Echinoderm mathaei*, using tetracycline. In *Echinoderm biology. Proc. 6th Int. Echinoderm Conf*, ed. R. D. Burke, P. V. Mladenov, P. Lambert, and R. L. Parsley, 435–43. Rotterdam: A. A. Balkema.

———. 1999. *Plant and animal populations. Methods in demography*. San Diego, CA: Academic Press.

Ebert, T.A., and M.P. Russell. 1993. Growth and mortality of subtidal red sea urchins (*Strongylocentrotus franciscanus*) at San Nicolas Island, California, USA: problems with models. *Mar. Biol.* 117:79–89.

Foster, M.S., T.A. Dean, and L.E. Deysher. 1985. Subtidal techniques. In *Handbook of phycological methods. Ecological field methods: Macroalgae,* ed. M.M. Littler and D.S. Littler, 199–231. Cambridge: Cambridge University Press.

Frank, P.W. 1965. Shell growth in a natural population of the turban snail, *Tegula funebralis*. *Growth* 29:395–403.

———. 1975. Latitudinal variation in the life history features of the black turban snail *Tegula funebralis* (Prosobranchia: Trochidae). *Mar. Biol.* 31:181–92.

Fretter, V., and A. Graham. 1964. Reproduction. In *Physiology of Mollusca, Vol.* 1, ed. K. M. Wilbur and C. M. Yonge, 127–64. New York: Academic Press.

Garbary, D.J., and R.E. DeWreede. 1988. Life history phases in natural populations of Gigartinaceae (Rhodophyta): quantification using resorcinol. In *Experimental phycology. A laboratory manual*, ed. C. S. Lobban, D. J. Chapman, and B. P. Kremer 174–78. Cambridge: Cambridge University Press.

Gosselin, L.A. 1993. A method for marking small juvenile gastropods. *J. Mar. Biol. Assoc. U.K.* 73:963–66.

Grant, A., P.J. Morgan, and P.J.W. Olive. 1987. Use made in marine ecology of methods for estimating demographic parameters from size/frequency data. *Mar. Biol.* 95:201–8.

Gunnill, F.C. 1980. Demography of the intertidal brown alga *Pelvetia fastigiata* in southern California, USA. *Mar. Biol.* 59:169–79.

Harper, J.L. 1977. *Population biology of plants*. London: Academic Press.

Hutchings, M.J. 1986. Plant population biology. In *Methods in plant ecology*. 2nd ed. P.D. Poore and S.B. Chapman, 377–435. Oxford: Blackwell Scientific.

Joule, B.J. 1983. An effective method for tagging marine polychaetes. *Can. J. Fish. Aquat. Sci.* 40:540–41.

Ju, S.-J., D.H. Secor, and H.R. Harvey. 1999. Use of extractable lipofuscin for age determination of blue crab *Callinectes sapidus*. *Mar. Ecol. Progr. Ser.* 185:171–79.

Kain, J.M. 1963. Aspects of the biology of *Laminaria hyperborea*. II. Age, weight, and length. *J. Mar. Biol. Assoc. U.K.* 43:129–51.

———. 1971. *Synopsis of biological data on* Laminaria hyperborea. FAO Fisheries Synopsis No. 87. Rome: FAO.

———. 1975. The biology of Laminaria hyperborea. VII. Reproduction of the sporophyte. *J. Mar. Biol. Assoc.U.K.* 55:567–82.

Kido, J.S., and S.N. Murray. 2003. Variations in owl limpet *Lottia gigantea* population structures, growth rates, and gonadal production on southern California rocky shores. *Mar. Ecol. Progr. Ser.* 257:111–24.

Klinger, T., and R. E. DeWreede. 1988. Stipe rings, age, and size in populations of *Laminaria setchelli* Silva (Laminariales, Phaeophyta) in British Columbia, Canada. *Phycologia* 27:234–40.

Knight, M., and M. Parke. 1950. A biological study of *Fucus vesiculosus* L. and F. *serratus* L. *J. Mar. Biol. Assoc. U.K.* 29:439–514.

Krouse, J. S., and G. E. Nutting. 1990. Evaluation of coded microwire tags inserted in legs of small juvenile American lobsters. *Am. Fish. Soc. Symp.* 7:304–10.

Lindberg, D. R., and W. G. Wright. 1985. Patterns of sex change of the protandric patellacean limpet *Lottia gigantea* (Mollusca: Gastropoda). *Veliger* 27:261–65.

Lüning, K. 1979. Growth strategies of three *Laminaria* species (Phaeophyceae) inhabiting different depth zones in the sublittoral region of Helogland (North Sea). *Mar. Ecol. Progr. Ser.* 1:195–207.

Lutz, R. A., and D. C. Rhoades. 1980. Growth patterns within the molluscan shell: an overview. In *Skeletal growth in aquatic organisms. Biological records of environmental change*, ed. D. C. Rhoades and R. A. Lutz, 203–54. New York: Plenum Press.

Mann, K. H., and C. Mann. 1981. Problems of converting linear growth increments of kelps to estimates of biomass production. In *Proc. 10th Int. Seaweed Symp*, ed. T. Levring, 699–704. Berlin: De Gruyter.

Macdonald, P. D. M., and T. J. Pitcher. 1979. Age-groups from size-frequency data: a versatile and efficient method of analysing distribution mixtures. *J. Fish. Res. Bd. Can.* 36:987–1001.

Mathieson, A. C., and Z. Guo. 1992. Patterns of fucoid reproductive biomass allocation. *Br. Phycol. J.* 27:271–92.

Moran, A. L. 2000. Calcein as a marker in experimental studies newly-hatched gastropods. *Mar. Biol.* 137:893–98.

Murray, S. N., and P. S. Dixon. 1992. The Rhodophyta. Some aspects of their biology. III. *Oceanogr. Mar. Biol. Annu. Rev.* 30:1–148.

Nakahara, H. 1961. Determination of growth rates of the nacreous layer by the administration of tetracycline. *Bull. Natl. Pearl Res. Lab.* 6:607–14.

Nielson, L. A. 1992. *Methods of marking fish and shellfish*. Am. Fish. Soc. Spec. Publ. 23.

Niemeck, R. A., and A. C. Mathieson. 1976. An ecological study of *Fucus spiralis* L. *J. Exp. Mar. Biol. Ecol.* 24:33–48.

Osenberg, C. W., R. J. Schmitt, S. J. Holbrook, K. E. Abu-Saba, and A. Russell Flegal. 1996. Detection of environmental impacts. In *Detecting ecological impacts. Concepts and applications in coastal habitats*, ed. R. J. Schmitt and C. W. Osenberg, 83–108. San Diego, CA: Academic Press.

Paine, R. T., C. J. Slocum, and D. O. Duggins. 1979. Growth and longevity in the crustose red alga *Petrocelis middendorfii*. *Mar. Biol.* 51:185–92.

Palmer, A. R. 1982. Growth in marine gastropods: a non-destructive technique for independently measuring shell and body weight. *Malacologia* 23:63–73.

Parke, M. 1948. Studies on the British Laminariaceae: 1. Growth in *Laminaria saccharina* (L.) Lamour. *J. Mar. Biol. Assoc. U.K.* 17:652–709.

Pirker, J. G., and D. R. Schiel. 1993. Tetracycline as a fluorescent shell-marker in the abalone *Haliotis iris*. *Mar. Biol.* 116:81–86.

Powell, J. 1986. A short day photoperiodic response in *Constantinea subulifera*. *Am. Zool.* 26:479–87.

Robertson, B. L. 1987. Reproductive ecology and canopy structure of *Fucus spiralis* L. *Bot. Mar.* 30:475–82.

Rosenthal, R. J., W. D. Clarke, and P. K. Dayton. 1974. Ecology and natural history of a stand of giant kelp, *Macrocystis pyrifera* off Del Mar, California. *Fish. Bull.* 72:670–84.

Rowley, R. J., and D. J. Mackinnon. 1995. Use of the fluorescent marker calcein in biomineralization studies of brachiopods and other marine organisms. *Bull. Inst. Océanogr. Monaco* (Num. Spéc. 14) 2:111–20.

Russell, G. 1979. Heavy receptacles in estuarine *Fucus vesiculosus* L. *Estuarine Coastal Mar. Sci.* 9:659–61.

Santelices, B. 1990. Patterns of reproduction, dispersal and recruitment in seaweeds. *Oceanogr. Mar. Biol. Annu. Rev.* 28:177–276.

Seapy, R. R. 1966. Reproduction and growth in the file limpet, *Acmaea limatula* Carpenter, 1864 (Mollusca: Gastropoda). *Veliger* 8:300–10.

Seed, R. 1980. Shell growth and form in the Bivalvia. In *Skeletal growth in aquatic organisms. Biological records of environmental change*, ed. D. C. Rhoades and R. A. Lutz, 23–67. New York: Plenum Press.

Sire, J. Y., and P. Bonnet. 1984. Croissance et structure de l'opercule calcifie du gasteropode polynesien *Turbo setosus* (Prosobranchia: Turbinidae): determination de l'age individuel. *Mar. Biol.* 79:75–87.

Subrahmanyan, R. 1960. Ecological studies on the Fucales I. *Pelvetia canaliculata* Dec. et Thur. *J. Indian Bot. Soc.* 39:614–30.

———. 1961. Ecological studies on the Fucales II. *Fucus spiralis* L. *J. Indian Bot. Soc.* 40:335–54.

Uglem, J., and S. Grimson. 1995. Tag retention and survival of juvenile lobsters, *Homarus americanus* (L.) marked with coded wire tags. *Aquacult. Res.* 26:837–41.

Van Montfrans, J., J. Capelli, R. J. Orth, and C. H. Ryer. 1986. Use of microwire tags for tagging juvenile blue crabs (*Callinectes sapidus* Rathbun). *J. Crustac. Biol.* 6:370–76.

Vermeij, G. J. 1980. Gastropod shell growth rate, allometry, and adult size: environmental implications. In *Skeletal growth in aquatic organisms. Biological records of environmental change*, ed. D. C. Rhoades and R. A. Lutz, 379–94. New York: Plenum Press.

Vernet, P., and J. L. Harper. 1980. The cost of sex in seaweeds. *Biol. J. Linn. Soc.* 14:129–38.

Werner, P. A., and H. Caswell. 1977. Population growth rates and age versus stage-distribution models for teasel (*Dipsacus sylvestris* Huds.). *Ecology* 58: 1103–11.

Wilbur, K. M., and G. Owen. 1964. Growth. In *Physiology of Mollusca, Vol. 1*, ed. K. M. Wilbur and C. M. Yonge, 211–42. New York: Academic Press.

Wright, W. G., and D. R. Lindberg. 1979. A non-fatal method of sex determination for patellacean gastropods. *J. Mar. Biol. Assoc. U.K.* 59:803.

Zar, J. H. 1999. *Biostatistical analysis*. 4th ed. Upper Saddle River, NJ: Prentice–Hall.

SUBJECT INDEX

TAXONOMIC INDEX

ABOUT THE AUTHORS

DR. STEVEN N. MURRAY

Steve Murray is Dean of the College of Natural Sciences and Mathematics and Professor of Biology at California State University, Fullerton. He received his B.A. and M.A. degrees from the University of California (UC), Santa Barbara, and his Ph.D. from UC Irvine. For his dissertation work, he studied the reproduction and life histories of red seaweeds under the mentorship of Peter Dixon. He has been studying coastal communities and processes for more than 30 years as a faculty member at California State University, Fullerton. Steve has published on topics such as marine herbivory, biogeography, human impacts on coastal populations and communities, genetic structure of coastal populations, and physiological ecology and reproductive biology of seaweeds. Most of his research has been performed in temperate California waters. His recent research has focused on determining long-term changes in species distributions and abundances, productivity, and food web structure in rocky intertidal communities over a period of shifts in oceanographic climate and increasing urbanization. Steve's research has included evaluation of the success of marine protected areas (MPAs) in protecting rocky intertidal populations and communities receiving high amounts of human use. This research has involved assessments of the abundances and structures of extracted and trampled populations inside and outside of long-standing coastal MPAs. He is also participating in a program designed to monitor temporal changes in rocky intertidal populations and communities over regional spatial scales. Recently, Steve's research has included studies of introduced seaweeds to determine their rates of spread, habitat utilization, and physiological performances.

DR. RICHARD F. AMBROSE

Rich Ambrose is a professor at the Department of Environmental Health Sciences and Director of the Environmental Science and Engineering Program at the University of California, Los Angeles (UCLA). He received his B.S. in Biological Sciences from UC Irvine and his Ph.D. in Marine Ecology from UCLA. After conducting postdoctoral research at Simon Fraser University in Vancouver, British Columbia, he spent seven years at the Marine Science Institute at UC Santa Barbara before coming to UCLA. Rich's research focuses on the ecological aspects of environmental problems, particularly in coastal environments, and he has worked in intertidal, subtidal, and wetland habitats. His current research on environmental issues centers on the restoration of degraded coastal habitats, especially wetlands, and ecological monitoring and assessment. He is currently working on several tidal wetland restoration projects in California. Other research projects include monitoring change in rocky intertidal habitats (with a particular focus on being able to detect effects of oil spills) using a network of monitoring sites throughout southern California, assessing the nature of human activities in rocky intertidal habitats and determining ways to restore degraded intertidal communities, assessing the impacts of contaminants on coastal wetland species, developing performance standards for determining the success of habitat restoration projects, understanding the dynamics of sea urchin-dominated rocky reefs, and evaluating alternatives for managing watershed-level ecological problems.

DR. MEGAN N. DETHIER

Megan Dethier grew up spending summers on the shores of New England and was thus pre-adapted to become a marine biologist. She did her undergraduate work at Carleton College in Minnesota, despite the apparent lack of ocean there, and then did her Ph.D. work under Bob Paine at the University of Washington, near a real ocean. Her dissertation revolved around the community ecology of intertidal pools. Since completing her graduate work, Megan has been in residence at the Friday Harbor Laboratories and is a Research Associate Professor at the Zoology Department of University of Washington. She has thus worked on shoreline ecology of the Pacific Northwest for over 25 years, first exclusively on rocky shores but now also in mud, gravel, and salt marsh habitats. She designed a marine habitat classification system for Washington state and has helped the National Park Service and various Washington state agencies design shoreline mapping and monitoring programs. Megan's current research efforts include investigating the linkage between physical features of shoreline habitats and their biota, studying the plant/herbivore ecology and ecophysiology of an intertidal seaweed, and investigating interactions between native salt marsh communities and an invasive cordgrass in Puget Sound.

Interior Design:	Jessica Grunwald
Composition:	TechBooks, Inc.
Text:	10/12 Baskerville
Display:	Baskerville
Printer and Binder:	Friesens Corporation